TELLING GENES

Telling Genes
The Story of Genetic Counseling in America

ALEXANDRA MINNA STERN

The Johns Hopkins University Press
Baltimore

The Johns Hopkins University Press
2715 North Charles Street
Baltimore, Maryland 21218-4363
www.press.jhu.edu

Library of Congress Cataloging-in-Publication Data
Stern, Alexandra.
    Telling genes : the story of genetic counseling in America / Alexandra Minna Stern.
        p. ; cm.
    Includes bibliographical references and index.
    ISBN 978-1-4214-0667-1 (hdbk. : alk. paper) — ISBN 1-4214-0667-5 (hdbk. : alk.
paper) — ISBN 978-1-4214-0668-8 (pbk. : alk. paper) — ISBN 1-4214-0668-3 (pbk. :
alk. paper) — ISBN 978-1-4214-0748-7 (electronic) — ISBN 1-4214-0748-5 (electronic)
    I. Title.
    [DNLM: 1. Genetic Counseling—United States. 2. Genetic Counseling—history—
United States. QZ 50]
    362.196′042—dc23   2012002232

A catalog record for this book is available from the British Library.

*Special discounts are available for bulk purchases of this book. For more information,
please contact Special Sales at 410-516-6936 or specialsales@press.jhu.edu.*

The Johns Hopkins University Press uses environmentally friendly book materials,
including recycled text paper that is composed of at least 30 percent post-consumer
waste, whenever possible.

CONTENTS

ACKNOWLEDGMENTS

A National Endowment for the Humanities Fellowship and a Publication Grant from the National Library of Medicine at the National Institutes of Health generously supported the research and writing of this book.

An earlier version of chapter 5 was published as "A Quiet Revolution: The Birth of the Genetic Counselor at Sarah Lawrence College, 1969," *Journal of Genetic Counseling* 18, no. 1 (2009): 1–11.

I am grateful to the many archivists who assisted me with this project, often identifying related or recently accessioned materials, all of which have greatly enhanced this book. They include Abby Lester at the Sarah Lawrence College Archives, Beth Kaplan at the University Archives of the University of Minnesota Libraries, Earle Spamer at the American Philosophical Society Library, David Rose at the March of Dimes Archives, Erika Gorder at the Rutgers University Archives, Phoebe Letocha at the Alan Mason Chesney Medical Archives of the Johns Hopkins Medical Institutions, Monica Garrett at the Dorothy Carpenter Medical Archives of the Wake Forest Baptist Medical Center, and Karen Jania and Malgosia Myc at the Bentley Historical Archives of the University of Michigan. I also thank the Bentley's director, Fran Blouin, for providing sagacious guidance on my Institutional Review Board application to consult the University of Michigan Adult Medical Genetics Clinic Records. In addition, this book has benefited from the personal archives, letters, and brochures that many people shared with me along the way, above all Margie Goldstein, Robert Resta, and Lucille Poskanzer.

The heart and soul of this narrative are the genetic health professionals that I interviewed (a full list of interviewees is included in appendix B). I hope I have captured their voices and experiences accurately and respectfully. Each interviewee was given the opportunity to review the chapters in which she or he was cited or referenced, and the comments I received during this process have strengthened the book. Deep thanks to Robert Resta, who went far beyond the call of duty by scrutinizing many iterations of the manuscript and putting me in contact with key figures in the world of genetic counseling. In addition, I am indebted to Barbara Biesecker, Marian Rivas, Virginia Corson, Judith Tsipis, Luba Djurdjinovic, Jon Weil, Dorene Markel, Lucille Poskanzer, Joan Marks, and Wendy Uhlmann for the extensive time and energy they put into reading and rereading sections of the manuscript in progress.

One of the most exciting aspects of undertaking a big scholarly project is presenting new material and analysis to colleagues at different stages. Lectures, conference panels, and workshops provided the opportunity for invaluable feedback, and I thank the following organizations and institutions for allowing me to share my work with them: the Department of Human Genetics, Genetic Counseling Program, ELSI Personal Genomics Seminar Series, and Program in Science, Technology, and Society at the University of Michigan; the Program in the History of Science, Medicine, and Technology at the Johns Hopkins University; the Department of History at the University of Windsor; the National Society of Genetic Counselors; the National Human Genome Research Institute/Johns Hopkins University Genetic Counseling Training Program; the Division of Medical Genetics, Department of Bioethics and Humanities, and Center for Genomics and Healthcare Equality at the University of Washington; the Department of Medical Genetics, Genetic Counseling Program, and Department of Medical History and Bioethics at the University of Wisconsin at Madison; the Program in Science in Human Culture at Northwestern University; and the Department of History at the University of South Florida.

The Center for the History of Medicine at the University of Michigan has been an engaging and inspiring place to conceive and complete this project, thanks largely to its director, Howard Markel, and his vast knowledge of and insights on the history of science and medicine. Over the past several years colleagues and friends have taken time away from their demanding schedules to read parts of the manuscript, and their input has enhanced the book. In particular, I would like to thank Nathaniel Comfort, a kindred traveler in the history of medical genetics, as well as Diane Paul, Martin Pernick, Paul Edwards, Gabrielle Hecht, Laura Hirshbein, Michelle McClellan, Elizabeth Roberts, Adam Warren, Barbara Berglund, and Mary Beth Reilly. In addition, Steve Epstein, Sue Lederer, Judy Leavitt, Wylie Burke, Alison Bashford, Kathleen Canning, Rayna Rapp, Steve Palmer, Jennifer Robertson, Celeste Brusati, Chris Hodgkinson, Steve Luxenberg, Molly Ladd-Taylor, Philip Giampietro,

Shobita Parthasarathy, and Ed Goldman all asked me challenging questions along the way that helped me clarify key points in the narrative.

It has been a joy to work with Jackie Wehmueller at the Johns Hopkins University Press. Her graceful combination of enthusiasm and incisive criticism has made her the ideal editor. I also thank Sara Cleary for her assistance during the review and production phases.

Terri Koreck has supported this project and taken up the slack at home when I was in intensive writing mode or out of town for research or travel, and I thank her for being a wonderful and loving partner. Our daughter, Sofia, also has had to endure the time demands of this book, and she suggested recently that "I write a small book next time, like *Norman the Doorman*"—a smart idea indeed.

Finally, I thank my father, Andrew Stern, for his ongoing love, and dedicate this book to my late mother, Mary Lou Wyatt Stern, whom I miss tremendously.

Introduction

When they visit their doctor, read the newspaper, or embark on the journey of pregnancy or parenting, Americans encounter genomic medicine. How they receive information about their own personal genetic makeup—and from whom—is immensely important to their lives and futures. Individuals and families navigating human genomics in the health care arena include pregnant women deciding whether to pursue prenatal genetic screening, perhaps followed by diagnostic procedures such as amniocentesis or chorionic villus sampling (CVS); persons at risk of incurable inherited disorders such as Huntington's disease who have the wrenching task of deciding whether to undergo genetic testing; and women whose biological relatives have hereditary breast cancer and who are facing tough decisions about medical management, including the possibility of prophylactic mastectomy and oophorectomy.

For more than 60 years, genetic counselors have served as the messengers of important information about the risks, realities, and perceptions of genetic conditions. Ideally, genetic counselors reliably translate test results and technical language for a diverse clientele, aiding them with equal doses of scientific acumen and human empathy to make decisions about their options. Genetic counselors can help to ensure that bioethical principles, such as patient autonomy and informed consent, are followed in clinical genetics as they are in other clinical specialties. These unique health care professionals play a key role in educating physicians, scientific researchers, and the lay public about the role of genetics in rare diseases and, increasingly, common and chronic conditions.

As of 2012, there are more than 3,000 certified genetic counselors in the United States, working in settings that range from county clinics, community and teaching hospitals, public health departments, and private biotech companies to academic medicine. Today most genetic counselors are board certified and have graduated from two-year master's programs that require courses in genetics and psychology, as well as clinical and research field placements. Most certified genetic counselors believe strongly that such potentially life-altering information ought to be communicated in a setting of client-centered care, which has become a central tenet of board-certified genetic counseling. Many other health care professionals, including nurses, social workers, and physicians from myriad specialties, communicate genetic risk information or test results and often seek to incorporate genetic counseling's core principles into their approaches.

How did the field of genetic counseling evolve? How have genetic counselors influenced and been influenced by developments in science, medicine, technology, and society? How has genetic counseling helped to frame understandings and experiences of genomic medicine in modern America? Where does genetic counseling fit on the biomedical landscape and as a helping profession with a particular brand of psychological and clinical training? What distinguishes interactions between genetic counselors and patients or clients, and how have these interactions changed over time? To explore these questions, this book places contemporary genetic counseling in its historical context and considers its role in the twenty-first century. I trace the origins of genetic counseling to the eugenics movement of the early 1900s, examine the field's emergence as an integral component of medical genetics in midcentury, and describe the convergence of scientific and social factors in the 1960s and 1970s which gave birth to the genetic counselor as we know her— and, to a lesser extent, him—today.

Over the past six decades, genetic counseling has changed dramatically. From giving advice aimed largely at influencing the genetic composition of the entire living population and future generations, genetic counselors now focus on the immediate and longer-term welfare of the individual and her or his family. This transition did not occur in clearcut chapters labeled "Before" and "After" but rather has involved a protracted and uneven process in which many pressing ethical issues related to reproduction, genetics, stigma, and difference remain unresolved. To a great extent the lack of resolution is an inevitable by-product of being a helping profession at the center of human genomics, a fast-moving area of scientific inquiry and research that generates new knowledge, technologies, and treatments related to diseases, rare and common, which come with profound social, ethical, legal, and personal ramifications.

In this book I demonstrate why history matters to contemporary genetic counseling practices, principles, and professionals. The field of genetic counseling carries burdensome historical baggage that imposes limitations and can unwittingly hinder the field and its practitioners. To begin with, according to the most recent professional survey, 95 percent of genetic counselors are female and 92 percent are classified as white or Caucasian. Clearly, members of the profession do not represent the racial, ethnic, and gender makeup of their clients and society at large, which raises serious questions about the ability of genetic counselors, no matter how well trained or intentioned, to reach and communicate effectively with clients from differing social and cultural backgrounds.[1] Furthermore, when genetic counseling emerged in the 1940s, many of its leading practitioners had no qualms about supporting the sterilization of people with developmental disabilities. Until at least the 1980s, many genetic counselors who strongly endorsed patient autonomy and informed consent nonetheless tended to present genetic conditions likely to produce a physical or mental disability as biological errors to be avoided for medical, psychological, and economic reasons. Twenty-first-century genetic counselors are worlds away from the eugenicists of yesteryear, and the recent generations of practitioners often have been at the forefront of introducing bioethical principles into the delivery of genetic services. Nevertheless, the profession has palpable links to the ugly side of hereditarian thinking, and these associations merit research and analysis.

Uncovering the acknowledged and unacknowledged continuities between the eugenic past and genomic present can provide insight into the genesis of popular attitudes, both negative and positive, toward genetic counselors and, more broadly, can provide guideposts for health professionals seeking to address the potential or pressing moral dilemmas attached to genetic technologies. The evolution of genetic counseling also sheds light on how a small yet sturdy health profession can mature swiftly, propelled by discoveries in clinical medicine and laboratory science and influenced by far-reaching economic, social, and cultural developments in American society.

Using historical methods, I narrate the story of genetic counseling in America from a perspective that is simultaneously critical and sympathetic. I plot the twists and turns over time by combining the paper trail from the archives with the voices of genetic health practitioners. The extensive research I carried out for this book involved many hours in historical repositories of universities, in special collections of medical schools, and with patient and kindred files. In addition, because genetic counseling has a vibrant living history, this book draws amply from

46 oral history interviews I conducted with genetic counselors, medical geneticists, and psychologists.[2] These complementary methods and sources illustrate that the development of genetic counseling was more circuitous than linear, influenced as much by scientific advances as by moral and emotional concerns about the consequences of genetic knowledge and disease.

Moving backward from today, chapter 1 situates the development and definitions of genetic counseling from the 1940s to the present. It begins with the moving story of Judith Tsipis, whose family history and scientific expertise compelled her to establish a master's degree genetic counseling program at Brandeis University in 1992. This chapter then traces the coining of the term "genetic counseling" by the University of Minnesota's Sheldon Reed in 1947 and analyzes the strong albeit ambivalent relationship of early genetic counselors to eugenic ideas and organizations. It provides a framework for understanding the dramatic transformation that occurred in the early 1970s when a new generation of women and men launched professional genetic counseling master's programs.

In chapter 2, I consider genetic risk, one of the cornerstones of genetic counseling, from the perspective of both practitioners of genetic counseling and clients. I offer background on the historical development of the concept of risk in the modern era and specifically trace the historical genealogies of the concept of genetic risk. By incorporating patient stories and experiences, this chapter shows how clients and patients experienced risk in different ways. For many clients, the mere possibility of any risk justifies genetic screening or testing, a trend that has accelerated as genetic risk calculations have become more accurate. The friction between subjective understandings of risk and genetic counselors' attempts to standardize risk figures is pivotal to grasping the unique professional role of the genetic counselor and social attitudes toward genetic counseling in the twentieth and twenty-first centuries.

Chapter 3 highlights the relationship between genetic counseling and understandings of race in science and society. Even as leading geneticists in the 1950s abandoned crude biological concepts of race, racialized differences continued to play a significant role in genetic counseling and medical genetics. For example, two of the country's heredity clinics made a Faustian bargain with the dogmatic benefactor of the racist Pioneer Fund in order to support research projects on mating patterns and human heredity. At the same time, geneticists affiliated with heredity clinics served regularly as the foremost experts in adoption cases when prospective parents or welfare agencies wanted a racial classification to match relinquished newborns with viable adoptive parents. Using a color scale devised by the biased British biologist Reginald Ruggles Gates, geneticists collaborated with colleagues in medicine and anthropology to categorize skin color, lip shape, hair texture, and other characteristics. The chapter explores this largely forgotten dimension of genetic counseling during the heyday of domestic adoption in America from the 1940s to the 1960s and discusses its relevance to issues of race and racial thinking in contemporary human genomics.

I extend the discussion of difference to disability in chapter 4, exploring the complex and seemingly contradictory ways in which medical geneticists and genetic counselors have simultaneously stigmatized disability and empowered persons with disabilities and their families. This chapter explores the role of genetic counselors in facilitating the creation of new biosocial identities and networks with and for people with disabilities. It also examines the reasoning of a group of human geneticists in the early 1960s that sought to replace the diagnostic labels of mongolism and Mongolian idiocy with Down syndrome in the name of greater scientific accuracy and the conscious de-racialization of a term with derogatory connotations. Highlighting stories of contemporary genetic counselors drawn to the field in part because of personal experiences with or awareness of people with disabilities and their families, this chapter considers the fractious but promising relationship between genetic counseling and disability rights.

In chapter 5, I recount what happened in the late 1960s and 1970s when a new generation, made up largely of determined and intrepid women, transformed genetic counseling and created the genetic counselor as we know her today. Through the lens of the first genetic counseling master's program, founded at Sarah Lawrence College in 1969, this chapter elucidates the gender dynamics and social, cultural, and economic factors that converged to remake genetic counseling. It also examines the motivations and key figures involved in the founding of sister programs in the 1970s, including at Rutgers University, the University of Colorado Health Sciences Center, and the University of California at Berkeley. Using the sociological concept of emotional labor, I explore the unique profile of contemporary genetic counseling as a feminized health care profession that combines scientific knowledge, empathic communication, and information delivery.

Chapter 6 examines how bioethics has influenced genetic counseling, with a focus on the tenets of autonomy and nondirectiveness. I consider four streams that contributed to the installation of bioethical principles in genetic counseling in the 1970s: Carl Rogers's theories of client- and person-centered counseling, Sheldon Reed's complex approaches to counseling and his preference for the term "client" over "patient," the centrality of genetic counseling and testing to the birth of bioethics in the 1960s and 1970s, and the incorporation of psychosocial theories of human behavior into genetic counseling training programs. This chapter suggests that the venerated ethos of nondirectiveness once provided a useful philosophical platform for genetic counselors to develop ethical and meaningful methods of counseling but over time has become an obstacle to flexible counselor-counselee interactions.

Chapter 7 follows threads from earlier chapters into the 1970s and 1980s by exploring the establishment of prenatal diagnosis and probing how genetic counselors approached reproductive choice and technologies during the introduction of amniocentesis and genetic screening. This chapter closely examines the activities and objectives of the Prenatal Diagnostic Center at the Johns Hopkins University Hospital, which launched one of the first amniocentesis clinics in America in 1969. It situates the emergence of a brave new era of reproductive choice in the context of struggles over abortion, unequal access to reproductive and genetic technologies, the growing professional profile of master's-degreed genetic counselors, the bioethical grail of reproductive autonomy, and persistent negative attitudes about children and people with disabilities.

The book's concluding chapter assesses the relevance of the history of genetic counseling to multifaceted genomic medicine in the twenty-first century, which is characterized by the rapid growth of direct-to-consumer genetic testing and the continual discovery of genetic factors in chronic and common diseases. The conclusion also looks toward the future and reflects on the opportunities and challenges that likely will test genetic counseling in the coming years.

History

Genetic Counseling Develops

In 1990, Judith Tsipis, a biology professor at Brandeis University, was having lunch with several female colleagues.[1] It was a heady time in human genetics, and Tsipis and her companions were well aware of impending developments in the field. That very year, the gargantuan federal endeavor of the Human Genome Project was announced to the public with calls to map the entire human genome in the name of scientific inquiry and lifesaving cures.[2] Eventually the Boston lunch discussion turned to recent breakthrough research in Huntington's disease (HD), notably the identification of a genetic marker on chromosome 4 associated with this devastating adult onset neurological disease. One of Tsipis's colleagues mentioned a recent article by Nancy Wexler, a leader in HD research, which expounded on the need for increasing numbers of genetic counselors who could address the psychological issues that would arise once HD testing became available, along with, more broadly, a growing battery of prenatal, newborn, and presymptomatic genetic tests.[3]

Over the course of the conversation, Tsipis became ardently convinced that she must start a genetic counseling program at Brandeis University in anticipation of the incipient explosion of genomic knowledge and technologies.[4] Since the founding of the first master's-level genetic counseling program at Sarah Lawrence College in 1969, 17 more programs had been established in the United States, 14 of which were still in existence (see appendix C for a chronological list of master's degree genetic counseling programs). Although Boston was home to a high concentration of renowned academic and medical institutions, the area lacked such a program.

Tsipis, a biologist with a doctorate from the Massachusetts Institute of Technology (MIT) and extensive training in bacterial and animal virus genetics, was fascinated by the scientific discoveries being made at the molecular level of DNA, but she was also drawn to human genetics for personal reasons. In 1975, Tsipis had given birth to a son, Andreas, who at 6 months began to exhibit developmental delays. Every year his condition worsened. He was seen by a legion of specialists, none of whom could provide an accurate diagnosis. Not until the age of 13 was Andreas diagnosed with Canavan disease (CD), a rare autosomal recessive disorder characterized by progressive destruction of the central nervous system. Named for Myrtelle Canavan, the Boston physician who identified it in 1931, CD is one of the leukodystrophies; in this case, owing to a genetically altered form of the enzyme aspartoacylase, the body is unable to effectively metabolize N-acetylaspartic acid, resulting in neurological and physical deterioration. Because it is a recessive disorder, the affected individual must receive two copies of the altered CD gene, one from each parent, to develop the disease (and one to be an unaffected carrier).[5] Following the basic principles of Mendelian inheritance, the child of two carriers of the altered CD gene has a 25 percent chance of being homozygous and developing the disease.[6] Although CD can occur in any ethnic group, it is more common among Ashkenazi Jews.

That fateful lunch served as the catalyst for Tsipis to, as she said, "take my personal and professional lives and combine them to help other families like ours."[7] Thorough and tireless in her approach to most activities in life, Tsipis threw herself into learning everything she could about existing genetic counseling programs and mastering the administrative ropes to launch an applied master's degree program at Brandeis. She devoted endless hours of extra work, encountering some resistance from local and national genetic counselors who were not enthusiastic about her plan. Yet persistence paid off. In less than two years Tsipis had won over initially wary Brandeis administrators, laid the groundwork for clinical internships, outlined a curricular plan, and hired instructors to teach the required science and psychology courses.

In September 1992, Brandeis's program admitted its inaugural cohort of seven students and became the first genetic counseling program in New England. Distinguished by its emphasis on the experiences of people with disabilities and their families, Tsipis's achievement continues to this day. Poignantly, Andreas's legacy lives on in the program. Before his death in 1998, when he was 22 years old, Tsipis and her husband established a scholarship bearing his name to support genetic counseling students.[8]

In many ways, Tsipis's story exemplifies the development of contemporary genetic counseling. A smart, highly educated woman based in academia with access to significant resources, Tsipis was motivated by a mix of maternal love, parental anguish, scientific expertise, and people skills. Yet if her journey elucidates salient and emblematic themes in genetic counseling, it also represents just one of multiple pathways. There was and is no single route to the development of modern genetic counseling or to becoming a genetic counselor. Rather, the history of genetic counseling is one of overlapping trajectories that attracted and involved people from a wide array of professional and personal backgrounds. Their common denominator was being affected in ways large and small by human heredity and genetic disorders, and wanting to help clients and families understand and respond to the lived realities and future implications of genetic diseases and genomic information.

Most genetic counselors devote their time to obtaining and presenting test and screening results and then calculating and conveying genetic risk information to a wide range of clients, ideally in an empathetic and empowering manner. Genetic counselors are trained to present complex technical and scientific information in accessible language and to work with their clients to determine the most appropriate next step. There are myriad factors that make each session unique to the individual or family seeking consultation: the genetic disorder or diagnosis, the calculated level of risk, and the type of screening or diagnostic test—prenatal, newborn, or presymptomatic—as well as the client's gender, racial and ethnic background, and religious, cultural, moral, and personal values.[9]

As of 2012, there are 28 fully accredited genetic counseling master's programs in the United States and Canada, and more than 3,000 certified genetic counselors can be found across the spectrum of health care, in clinical, research, educational, and commercial settings.[10] According to a 2010 survey commissioned by the National Society of Genetic Counselors (NSGC), the greatest percentage of NSGC members, 37 percent, are employed at university medical centers, followed by 22 percent at private medical facilities and 16 percent at public hospitals. Genetic counselors are concentrated in prenatal and pediatrics clinics, in cancer genetics, and at specialty facilities focused on genetic disorders such as Marfan syndrome or HD. They also work in smaller numbers at diagnostic laboratories, for health maintenance organizations (HMOs), county health agencies, and pharmaceutical companies, and in field clinics.[11]

Over 90 percent of the genetic counselors surveyed by the NSGC in 2010 have earned master's degrees in genetic counseling or human/

medical genetics, while the remainder possess some combination of the degrees of BSN/RN, JD, MBA, MD, MPH, MSN, or PhD.[12] The American Board of Genetic Counseling (ABGC) certifies genetic counselors and accredits genetic counseling programs, and the NSGC publishes the field's flagship *Journal of Genetic Counseling*, convenes working committees, and organizes an annual conference and regional meetings.[13] Nevertheless, many health practitioners, including those without board certification or NSGC affiliation, participate more broadly in genetic counseling when they convey genetic information to patients and clients. Obstetricians regularly provide the results of amniocentesis, pediatricians work closely with children affected by genetic conditions and their families, ophthalmologists identify hereditary eye diseases and explain any available treatment options to patients, and oncologists discuss the availability of genetic tests for predispositions to breast, ovarian, or colorectal cancer.

Although there are many MDs who provide high-quality genetic counseling, overall physicians tend to be more prescriptive and less psychosocially oriented than trained genetic counselors. Furthermore, physicians are often pressured to see a high volume of patients in a short amount of time, making it difficult if not impossible to engage in an in-depth conversation. For example, the 2010 NSGC survey indicates that the majority (60%) of genetic counselors spend 31 to 60 minutes in face-to-face interaction with each patient, while the majority of physicians (52%) spend less than 15 minutes. This pattern can be a source of frustration to genetic counselors, but it can also help to justify their relevance as health care professionals who impart personalized care to patients and clients.[14]

Genomic medicine is exciting, is fast paced, and holds the promise of cures and prevention and the allure of biological optimization.[15] Every day we learn more about the genetic components of disease, when a rare genetic disorder is sequenced and mapped or when the relationship between genetic susceptibility and environmental factors for chronic diseases is further explicated.[16] According to the GeneTests database maintained by the National Institutes of Health (NIH), as of October 2011 there are tests for more than 2,440 genetic disorders and diseases, with close to 1,200 clinics providing such services and almost 600 laboratories processing specimens.[17] Genetic conditions are also less rare than many people think, increasing as contributors to morbidity and mortality in Westernized countries since the 1950s, after the introduction of broad-spectrum antibiotics and mass immunization programs effectively cured and treated many deadly infectious diseases. For example, today cystic fibrosis affects 1 in 2,500 Caucasian Americans, and 1 in 25 are carriers; 1 in 500 African American newborns are homozygous for the recessive disorder sickle cell anemia, while 1 in 12 are heterozygous carriers; and 1 in 8 American women will develop breast cancer, with 5–10 percent of these cases due to an inherited genetic predisposition.[18] In addition, Down syndrome, which 95 percent of the time is not an inherited genetic disorder, is the most common of one type of chromosomal anomaly, affecting every 1 in 600 to 700 births.[19] Adding to our burgeoning knowledge of genetics, scientists are discovering many of the genetic factors involved in common and chronic diseases, and the rapid amplification of genomic medicine has tremendous implications for genetic counseling and health care more generally.[20]

What Dorothy Nelkin and M. Susan Lindee wrote in *The DNA Mystique* over one decade ago rings true today: American society is awash in "genetic essentialism," ranging from the ubiquity of genetic explanations for a wide range of physical and psychological conditions to the belief that our authentic self can be revealed by our DNA.[21] Epidemiologist and feminist health scholar Abby Lippman employs the related term "geneticization" to describe the interplay of genetic determinism and the individualization of a wide swath of health and social problems through DNA discourse.[22] Genetic explanations and genomic research permeate clinical medicine, and genetic tests and counseling have reached far beyond the initial domains of prenatal and pediatric care to encompass ophthalmology, neurology, psychiatry, oncology, and dermatology, now woven into the fabric of biomedicine. We are overwhelmed with newspaper headlines announcing the discovery of genes for Alzheimer's disease, obesity, and gambling, and personalized genome kits costing $100 are targeted for marketing at major drugstore chains such as Walgreens.[23]

Unfortunately, most genetic knowledge is gained without the direct benefit of therapy or prevention.[24] Genetic tests might detect that a person possesses a genetic marker or predisposition for various genetic diseases, but in most cases little can be done except to treat secondary symptoms or wait anxiously for the full onset of the condition. Genetic and germ line therapies tantalize physicians and researchers but have delivered few—and sometimes fatal—results.[25] Indeed, one of the primary roles of genetic counselors is to help clients navigate the chasm between potentially life-changing genetic information and a limited armamentarium of medical therapies and options.[26]

For all these reasons, genetic counseling is more germane than ever in the twenty-first century. Yet as it moves forward, the field is weighed down by the past and jostled by the present. Decades-long historical legacies simultaneously bolster and impede genetic counseling in ways that might not be readily apparent but influence client consultation, attitudes about disability and abnormality, and the demographic profile of the profession. Most starkly, the shadow of eugenics hangs over the field. Critics contend that prenatal testing a priori equates disability with disease, and that the underlying assumption that the able-bodied are worth more than the disabled encourages the termination of pregnancies determined to be at higher risk for genetic diseases or with diagnosed chromosomal or anatomical anomalies.[27] Even if dissimilar to the state-sanctioned eugenics of the past, which entailed forced sterilization and marriage laws, the omnipresent pressure on American women to produce the "best" or healthiest child possible using available genetic and reproductive technologies resonates with the quest for superior biological fitness and could be considered neo-eugenic.[28] And while a minority of people can afford the latest genetic technologies, such as preimplantation genetic diagnosis (PGD), most cannot, potentially making the prevention of rarer genetic diseases a luxury for the wealthy few and unattainable for many.[29]

In addition to this historical baggage, genetic counseling is harshly challenged in today's health care environment. The genetic counselor is pulled and pushed in many directions and must keep pace with rapid developments in genomic medicine, the availability of new genetic technologies and tests, and the ever-growing knowledge of the genetic components underlying many debilitating and chronic diseases. Genetic counselors often find themselves in the unenviable position of trying to swim upstream, wanting to approach the client holistically as a complex individual but having no choice but to provide services in a health care system that is highly reductionist and privileges the disease over the patient. Two scholars wrote in 1995, as the Human Genome Project gained momentum, that "genetic counselors often experience a conflict between their role as empathetic nondirective supporter and role as information giver and decision-facilitator."[30] Although not eagerly discussed, genetic counseling sits at the crossroads of the abortion debate, with the potential to become particularly sensitive because terminations for a range of genetic conditions typically are performed during the latter two trimesters of pregnancy. Add to this cost cutting and the demands to see clients in shorter and shorter time frames, often with no promise of billable reimbursement, and it is remarkable that genetic counselors manage to have an affirming influence on the lives and decisions of patients and clients.

Given this combination of historical vicissitudes and contemporary crosscurrents, many genetic counselors today readily assert that their profession faces substantial obstacles and requires tough leadership, vision, and clout to survive and thrive.[31] Robin Grubs, codirector of the genetic counseling program at the University of Pittsburgh, expresses with regret that "little prestige, not much money, and a problematic history" characterize her profession.[32] Along with many of her colleagues who share a deep commitment to the field, Grubs is trying to change this predicament from both inside and out. For instance, genetic counseling programs are expanding their research tracks in order to measure the effectiveness of genetic counseling in terms of the satisfaction of clients and their families, to gauge the field's positive impact on decision making and medical management, and to document the economic benefits of genetic counseling to clinical care. Many genetic counselors serve on intramural and extramural boards and committees assessing the clinical, scientific, social, and ethical implications of genomic medicine. There is also a concerted push to bolster the profession by advocating for licensure of genetic counselors on the state level. As of 2011, only 11 states were issuing licenses in accordance with state law, although bills had recently passed or been introduced in committee in at least 10 more states.[33]

Genetic Counseling (Re)Defined

Given the velocity of developments in human genetics and the controversial issues attached to genetic testing, it is not surprising that the scope and mission of genetic counseling have been assessed multiple times since the field's emergence more than 60 years ago. Most recently, in 2003 the NSGC organized a committee to ascertain if the dominant definition of genetic counseling needed revision.[34] The impetus for this reevaluation was a concern that the foundational definition, put forth by the American Society of Human Genetics (ASHG) nearly 30 years earlier, was too verbose and, more important, out of sync with the myriad advances in genomic medicine that had altered the landscape of genetic counseling. Moreover, the NSGC leadership felt that it was high time for their organization, rather than the ASHG, which caters mainly to MDs and PhDs, to be the definer rather than the defined. After several months of deliberation, the NSGC's Genetic Counseling Definition Task Force proposed the following exposition:

> Genetic counseling is the process of helping people understand and adapt to the medical, psychological and familial implications of genetic contributions to disease. This process integrates the following:
> - Interpretation of family and medical histories to assess the chance of disease occurrence or recurrence.
> - Education about inheritance, testing, management, prevention, resources and research.
> - Counseling to promote informed choices and adaptation to the risk or condition.[35]

Approved in 2005 by the NSGC Board of Directors, this definition expressed the conscious choice to underline what genetic counseling *is* rather than what genetic counselors *do*, to distill the "essence of the relationship between the client and counselor when they interact."[36] In 2001, Barbara Biesecker, who served on the NSGC Genetic Counseling Task Force and is director of the genetic counseling program jointly run by the National Human Genome Research Institute at the NIH and the Johns Hopkins University, and colleague Kathryn Peters penned an influential definition of the field. The opening sentence of their concise version reads, "Genetic counseling is a dynamic *psychoeducational* process centered on genetic information"[37] (italics in the original). This and subsequent twenty-first-century definitions stress the interactive and psychosocial dynamic of genetic counseling, placing the emphasis on the therapeutic relationship between the counselor and the client, instead of simply listing the priority tasks performed by genetic counselors. According to Biesecker and Peters, the goal of the genetic counselor is to "facilitate clients' ability to use genetic information in a personally meaningful way that minimizes psychological distress and increases personal control."[38] These definitions suggest that genetic counseling sessions should consist of informed, educational consultations that are fluid and responsive to the genetic condition and the personal circumstances at hand.

Along these lines, contemporary definitions are attuned to the variety of scenarios encompassed by genetic counseling: for cancer genetics, where counselors would be remiss not to recommend a course of medical management; for predictive adult testing, where clients evaluate loaded ethical questions about confidentiality and the impact of sensitive genetic information on immediate and extended family members; and for prenatal screening, where counselors' overriding objective is to help parents make autonomous decisions about screening, diagnostic testing, possible preparation for the birth of a child with disabilities, or pregnancy termination.[39]

Such nuances were barely on the horizon three decades earlier when the ASHG formulated the first professionally sanctioned definition of genetic counseling. In 1970, the ASHG convened the Ad Hoc Committee on Genetic Counseling in an attempt to impose some order on a field that had existed amorphously since the 1940s but was undergoing transition and expansion in response to scientific developments and the burgeoning of master's-level genetic counseling programs. The Ad Hoc Committee on Genetic Counseling was faced with some important questions: Who was qualified to offer genetic counseling? Should the ASHG be responsible for the accreditation of genetic counselors? What were the immediate and long-term aims of genetic counseling?

In October 1971, this ad hoc committee issued an internal report tinged with alarmism that foresaw the marked growth of genetic counseling and worried about the ability of MDs and PhDs to maintain control over this budding field. The committee members concurred that genetic counseling should not become the domain of lesser-trained master's-degreed professionals, declaring that genetic counseling should be provided only by "physicians who are appropriately trained in genetics."[40] The 1971 report also urged the creation of a body charged with accreditation to ensure competence in clinical medicine and human genetics. Despite its proprietary proclamations, this report did little to resolve the issue for the ASHG. Its concerns still pending, the following year the ASHG reactivated the Ad Hoc Committee and appointed Charles J. Epstein, a pediatrician at the University of California at San Francisco (UCSF), to chair this revamped body. Epstein invited some of the country's most prominent geneticists, such as Curt Stern, Arno G. Motulsky, and F. Clarke Fraser, to join the deliberations.[41] The sense was that the committee needed to act quickly because new technologies, such as recombinant DNA, were remaking medical genetics at the molecular level, and that whether they liked it or not, genetic counseling's center of gravity would soon sit in master's degree programs. As one committee member exhorted to Epstein in 1972, we must be "clearheaded and acknowledge that the movement towards establishment of genetic counseling as a profession is not going to halt merely because of [*sic*] the ASHG can not afford to study it systematically at this time. Others are moving ahead."[42]

When he assumed the chair, Epstein firmly believed that only MDs and PhDs should practice genetic counseling, the chief goal of which was to prevent disease by communicating accurate risk information.[43] Reflecting with hindsight, Epstein explained that, at the time, he and many of his colleagues were worried about the correct "vetting of genetic information" and the potential for untrained counselors to "get into dangerous territory."[44] In October 1972, Epstein voiced these concerns at the ASHG meeting in a lecture that had immediate reverberations. Nearly 40 years later, fellow geneticist Kurt Hirschhorn still vividly remembers when Epstein stated in strong terms that non-MDs and non-PhDs should be excluded from genetic counseling, a position that pleased some but offended just as many.[45] For example, Marian Rivas, who had just assumed the directorship of the newly created genetic counseling program at Rutgers University, was caught off guard by the baldness of Epstein's remarks. A medical geneticist who had earned her PhD in a decidedly clinically oriented program at Indiana University, Rivas was keenly aware that two years was a short amount of time for students to grasp the complexities of genetic science and the diagnosis of hereditary diseases. Yet, after setting up a well-rounded curriculum, laboratory training, and clinical rotations for the Rutgers program and seeing firsthand the high caliber of the entering student cohort, Rivas was sanguine about the abilities of well-trained master's-degreed genetic counselors. Sitting in the audience during Epstein's talk, Rivas, who had brought her students to the ASHG to spur their professionalization, was crestfallen. Today, she recalls the distinct feeling of her heart sinking as Epstein talked. She looked around the room at her hardworking acolytes and wondered how she would ever buoy their spirits.[46]

Epstein's lecture, as well as the animated response to it, had a far-reaching impact on the definition and direction of genetic counseling. Receptive to the criticism he received during and after the ASHG meeting from colleagues he held in high regard, Epstein invited additional members such as Rivas and Margaret Thompson to join the Ad Hoc

Committee. Over the next three years, this group debated the meaning and scope of genetic counseling, eventually abandoning Epstein's original position. Against the backdrop of the successful graduation and job placement of the inaugural classes of genetic counselors trained at master's-level programs, and given a demonstrated need to staff proliferating genetic screening programs and prenatal clinics, many initially reticent medical geneticists became convinced that freshly minted genetic counselors could fill a valuable niche. Most strikingly, as Epstein observed the skills of the University of California at Berkeley genetic counseling students in UCSF's clinics, he altered his stance. He realized that his fears about master's-level genetic counselors delivering incorrect diagnoses and genetic information were unfounded, and by the late 1970s, Epstein had made a "180 degree turn" and become a vocal champion of the contemporary genetic counselor and an inspiring mentor to countless students.[47]

Epstein's and the ASHG's reversal was evident in the definition of genetic counseling its members approved in 1974, which became the norm until NSGC's 2005 revision:

> Genetic counseling is a communication process which deals with the human problems associated with the occurrence, or the risk of occurrence, of a genetic disorder in a family. The process involves an attempt by one or more appropriately trained persons to help the individual or family to (1) comprehend the medical facts, including the diagnosis, probable course of the disorder, and the available management; (2) appreciate the way heredity contributes to the disorder, and the risk of recurrence in specified relatives; (3) understand the alternatives for dealing with the risk of recurrence; (4) choose the course of action which seems to them appropriate in view of their risk, their family goals, and their ethical and religious standards, and to act in accordance with that decision; and (5) to make the best possible adjustment to the disorder in an affected family member and/or to the risk of recurrence of that disorder.[48]

In the end, the 1974 ASHG definition represented a significant departure from previous understandings of genetic counseling. Regina Kenen, a sociologist of professions, has insightfully noted that "in the mid-1970s a shift began to take place" as "concern with prevention became tempered by a concern with the client's total well-being" and "medical geneticists conducting counseling began to realize that preventing all or even most genetic defects might be an unattainable goal for the foreseeable and probable long-term future."[49] Now, in addition to correct diagnosis, medical knowledge, and risk assessment, a premium was placed on the interactive and communicative experience between the genetic counselor and the client.[50] The ASHG also modified its position in terms of practitioners, asserting that medical social workers, public health nurses, clinical psychologists, and, notably, genetic associates—as genetic counselors were briefly called—with adequate training could and should proffer genetic counseling.[51]

For the final quarter of the twentieth century the ASHG's definition provided a sturdy framework for genetic counselors. It heightened the profile of the field and "helped establish the acceptance of medical genetics and genetic counseling by the larger health care community."[52] One of the many ironies in the history of genetic counseling is that although the ASHG formulated the first working definition of genetic counseling in response to the gatekeeping anxieties of MD and PhD geneticists, by the time the heated discussions had ended and the ink had dried, human genetics no longer "owned" genetic counseling. By the 1970s, genetic counseling was becoming the province of a new generation of primarily professional women intent on the success of master's-level genetic counseling programs and the placement of their graduates in clinical and research settings.

The (Eu)genesis of Genetic Counseling

The ASHG deliberated the meaning of genetic counseling during a decade that saw major advances in molecular biology and genetic engineering and the introduction of a core set of ethical principles in biomedicine. However, the term "genetic counseling" predated this process by about three decades, first uttered during an era when the lines between medical genetics and eugenics were blurred. Historian of science Nathaniel Comfort has demonstrated that eugenics and human genetics have had a symbiotic relationship throughout the twentieth century, joined together by the twin and sometimes conflicting goals of disease prevention and the relief of human suffering.[53] Similarly, political scientist and well-known genetics scholar Diane Paul has argued that "throughout the 1940s, 1950s, and even the 1960s, few geneticists objected to the characterization of applied medical genetics as 'eugenics.'"[54] Genetic counseling emerged at the nexus of medical genetics and eugenics, even as its initial practitioners expressed great ambivalence over the possibilities for human betterment through genetic selection and the appropriate role of counselors in the domain of personal and family decision making about health and reproduction.

In any case, this was not the eugenics of the early twentieth century with its crude racial and ethnic prejudices; quick condemnation of poor, illiterate families such as Jukes or Kallikaks; and simplistic theory of monogenic Mendelian "unit characters."[55] The adherents of this revised, more democratic eugenics were post–World War II liberal scientists who publicly eschewed biological racism and believed that desirable and undesirable traits were distributed evenly across groups.[56] For the most part, they envisioned their clinical work, medical research, and educational programs as endeavors in harmony with liberal democracy. Often they were the only medical professionals to consult and console referred or walk-in patients whose lives and families had been upended physically and emotionally as a result of hereditary conditions.

But even if reformed, eugenics still had its biases, and they were in evidence to greater and lesser degrees at the country's first three human heredity clinics—the University of Minnesota's Dight Institute, the University of Michigan's Heredity Clinic, and Wake Forest University's Department of Medical Genetics—all founded in 1941.[57] Even though these clinics were distinct in terms of institutional arrangement and key personnel, they all were established on eugenic principles. The Dight Institute was founded thanks to a generous gift from Charles Dight, Minnesota's most famous eugenicist and the crusader behind the successful passage of that state's sterilization law in 1925.[58] The Heredity Clinic in Ann Arbor was largely the result of the efforts of Lee Raymond Dice, a zoologist specializing in field mice who became attracted to eugenics and human genetics in the late 1930s. In North Carolina, Wake Forest's department was launched by William Allan, a family doctor interested in genetic conditions who described the advice he gave patients suspected of hereditary defects as "voluntary negative eugenics."[59]

It was Sheldon Reed, director of the Dight Institute, who coined the term "genetic counseling" in 1947.[60] He attributed his neologism to the influence of Tage Kemp, a Danish geneticist practicing "genetic hygiene" in Copenhagen in the 1940s who wholeheartedly supported eugenic programs such as the sterilization of supposed hereditary defectives.[61] While Reed admired Kemp's work, he didn't like the word "hygiene," which he thought was "associated with tooth pastes, soap and other health care products" in the United States.[62] Struggling to find a new term, Reed settled on "genetic counseling," pinpointing the day he debuted this genetic health service as Monday, August 18, 1947.[63]

When Reed's pioneering work is discussed today, his definition of genetic counseling as a "kind of genetic social work without eugenic connotations" usually is invoked.[64] Yet genetic counselors do themselves a disservice by failing to acknowledge the midcentury (eu)genesis of the profession. Scratching the surface of Reed's long career, it quickly becomes clear that he did not disdain the aims of eugenics, but rather was not convinced that human genetics was yet able to supply the tools and knowledge to realize them.[65] In many of his papers, Reed expressed the viewpoint that eugenics was "the science of genetics applied to man" and that "human genetics and eugenics" were "synonyms."[66] He averred that once sufficiently advanced, medical genetics could supply the scientific facts and objectivity for an applied eugenics capable of improving the gene pool. Furthermore, Reed thought that if clients seeking advice made sound decisions about reproduction based on their family pedigrees, then genetic counseling could further human betterment. As he wrote, genetic counseling can play "a useful role in improving the genetics of the population because if a couple knows that they are expected to have no more than two or three children they will want to be sure that their two or three offspring are genetically normal."[67] In 1961 Reed wrote to Harry L. Shapiro, president of the American Eugenics Society (AES), to endorse the pursuit of both positive and negative eugenics because "man can improve his genotype eventually by conscious effort." Even though "we know very little yet as to how to set up precise eugenic goals," Reed thought that the AES "should devote its efforts toward this end."[68] As late as 1979, Reed explained in a lecture that "our present day use of the term 'human genetics' instead of 'eugenics' may be financially and politically expedient but there is no great philosophical distinction between them."[69]

However, in the 1970s, when Reed looked back at the more than 4,000 consultations performed at the Dight Institute during his directorship, he was unsure if the ultimate outcome of his counseling work had been dysgenic or eugenic, given that he usually encouraged his mainly middle-class educated clients to have more children even if their reproductive and family histories included probable congenital or inherited conditions.[70] And as often as Reed presented genetic counseling as commensurate with eugenics, at other times he described it as distinct from, even antithetical to, eugenics. For example, in 1972 Reed told his audience at the University of Nebraska that "the fact that there was an eventual genetic counseling explosion is partly due to my insistence from the beginning that genetic counseling has nothing to do, directly, with eugenics or any of the reproductive no-nos."[71] During these same decades, Reed also characterized genetic counseling as "investment counseling," practiced by an expert who "gives facts and predicts average expectations" in a "friendly, relaxed, and not threatening" environment.[72]

Reed's varying definitions are less evidence of duplicity or confusion than a reflection of the rich ambivalence about the purpose and scope of genetic counseling expressed among its self-appointed practitioners in midcentury America. From the 1940s to the 1960s, many of the prominent medical geneticists involved in genetic counseling—such as Reed, Dice, Laurence Snyder, Madge Macklin, and Nash Herndon—were members of both the AES and the ASHG, founded in 1948. For some, genetic counseling offered an opportunity to impose the "doctor knows best" mentality on the most intimate matters of human reproduction. At the same time, these early genetic counselors expressed a great deal of compassion for patients and their families, even as they stigmatized them as carriers of genetic defects. For example, Reed regularly received letters from parents who lavished him with praise for treating them with respect and not just as the progenitors of a freakish child.

The overlap between eugenics and medical genetics reached its zenith in the 1950s. *Eugenics Quarterly*, the journal of the AES, published a regular column on hereditary counseling from 1954 to 1958, a period that coincided with a climb in circulation from 281 to 1,033 subscribers.[73] In this special section, medical geneticists considered the relationship between genetic counseling and various eugenic ideas and policies. As Nash Herndon of Wake Forest University noted in 1953, both positive and negative eugenics were necessary since they each impacted gene frequencies. Following this logic, Herndon asserted that genetic counseling constituted applied eugenics: "As long as work with hereditary diseases remains purely investigative, it remains in the domain of the science of genetics, but as soon as we begin to apply our acquired knowledge to families or groups of people we are practicing eugenics."[74]

The first concerted effort to survey the purpose and scope of genetic counseling occurred in 1957 when the AES sponsored a heredity counseling symposium. A one-day event held at the New York Academy of Medicine, the conference was underwritten by the Rockefeller Foundation's Population Council, which paid "the expenses of those attending the meeting and a small honorarium."[75] It featured talks by all the leading lights in human genetics, including Reed, Fraser, Victor McKusick, and James Neel, among others. The conference participants, numbering more than 100, included many dual members of the AES and the ASHG, as well as family and marriage counselors, representatives of foundations involved in medical work, and physicians, nurses, and social workers.[76] After the closing session, the AES took advantage of the presence of so many of its members to hold its annual meeting.[77] Neel, only recently appointed inaugural chair of the Department of Human Genetics at the University of Michigan, told Osborn soon after the conference that he thought that the "symposium went off very well indeed."[78]

The proceedings from this conference were published both in the *Eugenics Quarterly* and in book form, outlining the concerns that motivated the conference and genetic counseling more generally.[79] These included the structure of heredity counseling services, the assessment and meaning of genetic risk, protocols for physician referrals, types of hereditary conditions, and comparisons with European clinics and methods. One of the most pressing questions was related to advice and patient interaction: does the genetic counselor "simply supply available information or should he advise the client as to his course of action?"[80] Participants fluctuated between seeing genetic counseling as a vehicle for eugenics and a psychological service that should respect the decision-making capacity of clients. Some believed that genetic counseling required heavy-handed recommendations in order to discourage procreation among couples likely to produce children with genetic defects, while others were adamant in their conviction that genetic counselors should avoid explicit advice or grandstanding. For example, psychiatrist Franz J. Kallmann explained that "persons requesting genetic advice cannot always be presumed to be capable of making a realistic decision as to the choice of a mate, or the advisability of parenthood, without support in the form of directive guidance and encouragement. As is true in recommending any restrictive regimen, persons coming to heredity clinics may have to be told *how* to adjust, for instance, to a childless marriage or a family deliberately limited in size"[81] (italics in original). Conversely, Reed told conference goers that the counseling offered to clients at heredity clinics "must be compassionate, clear, relaxed, and without a sales pitch."[82] From the 1940s to the mid-1960s, this generation of genetic counselors expressed varying views on the social and therapeutic aims of medical genetics, offering a spectrum of advice and attitudes that ranged from the overtly eugenic to the narrowly informative.

Enter the Genetic Counselor

While the first practitioners of genetic counseling were busy debating the reach and role of their expertise in people's reproductive lives, medical genetics was on a fast track. In the 1950s, chromatography for the screening of urine for abnormal metabolites became available, as did paper and starch gel electrophoresis of serum proteins, which in turn spurred interest in structural abnormalities in proteins and in hemoglobinopathies and other hereditary anemias. In 1956, Joe-Hin Tjio and Albert Levan demonstrated that the correct number of chromosomes in humans was 46, and three years later, Jérôme Lejeune found that an extra chromosome (trisomy 21) was the cause of Down syndrome. Over the next 10 years, researchers in Europe, North America, and Asia identified at least 100 chromosomal anomalies, including sex chromosome disorders such as Turner syndrome and Klinefelter syndrome.[83]

The medicalization of human genetics, which accelerated in the 1950s and 1960s, helped to break human genetics out of the container of organized eugenics, even as this process produced knowledge and technologies that raised and continue to raise profound philosophical questions about possibilities for both disease prevention and bioenhancement in genetic and reproductive health. As Lindee has pointed out, before the development of cytogenetics and karyotyping, genetic conditions could only be "seen or detected in the pedigree, the historical reconstruction of a family's history, and in the clinically abnormal body."[84] By the 1960s, a growing number of genetic conditions could be visualized and understood as the biochemical products of underlying disease process. Newborn testing for phenylketonuria (PKU), an inborn error of metabolism that results from a deficiency of phenylalanine hydroxylase, is characterized by mental retardation, and in many cases can be treated effectively with a low-phenylalanine diet initiated very early in infancy, was becoming standard practice throughout the United States.[85] In the 1970s, the concomitant emergence of amniocentesis and ultrasound technology spurred the steady expansion of prenatal diagnosis. The increasing availability of birth control, above all the pill, and the intensifying campaigns of the feminist health movement to make abortion accessible and legal were transforming the landscape of reproductive and genetic health.[86]

Onto this stage entered the contemporary genetic counselor. In 1969, the first cohort of genetic counseling master's students began their coursework and clinical training at a genetic counseling master's program at Sarah Lawrence College. By the early 1970s, similar programs had been founded at Rutgers University (Douglass College), the University of Pittsburgh, the University of California at Irvine, the University of Colorado Health Sciences Center in Denver, and the University of California at Berkeley, with more following by the decade's close. In 1979, after extensive legwork by genetic counselors and passionate wrangling about the naming and reach of the profession, the NSGC was created.[87] The American Board of Medical Genetics began certifying genetic counselors in 1982, an arrangement that lasted until that body decided to exclude master's-level genetic counselors from their purview. This "bitter divorce," as seasoned genetic counselor Robert Resta calls it, ultimately resulted in the ABGC becoming the designated credentialing body and had a sweet ending, allowing "genetic counselors to come from under the shadow of physician geneticists and be taken seriously as a separate profession."[88] Over the course of these three decades, genetic counselors became a small but critical component of biomedicine and health care, bolstered by the expansion of prenatal, newborn, and predictive genetic screening and testing in specialty clinics and in the offices of obstetricians, gynecologists, and primary care physicians. In addition, genetic counselors benefited from the growing availability of and demand for amniocentesis, and later chorionic villus sampling and maternal serum alpha-fetoprotein (MSAFP) screening, technologies encouraged among the growing number of older (over 35 years of age) mothers at greater risk of fetal chromosomal anomalies.[89]

Unlike many medical and health fields that evolve steadily over time, genetic counseling underwent a sea change in the 1970s, transforming from a field inhabited mainly by male experts under the auspices of heredity clinics and medical genetics departments to a profession populated principally by women with specialized master's degrees. For example, in the 1970s, when sociologist James Sorenson conducted the first systematic study of the nascent profession—its practitioners, principles, and practices—he counted approximately 650 men and women providing counseling on a regular basis at 285 centers around the country. The majority (72%) were men with an MD, and 64 percent of these were pediatricians, while the rest were primarily male scientists with PhDs.

Merely 7 percent held only master's degrees.[90] Little more than a decade later, there were approximately 700 master's-degreed genetic counselors providing the core of services and building the profession.[91] As Regina Kenen and Ann Smith have written, "It is relatively rare for a new occupational field to develop, expand, and change at such a rapid rate that a sociological analysis every decade provides illuminating insights into the internal and external factors fueling the shifts. The Masters Level Genetic Counselor is one of these rarities, elucidating in a time-reduced framework pressures felt and exerted by other occupational fields over longer time periods."[92]

Nevertheless, the remaking of genetic counseling in the 1970s did not mean that the field had liberated itself from a problematic past. Indeed, until the 1980s, it was not unusual for those involved in genetic counseling to espouse the logic of disease prevention for individual patients and the concomitant improvement of the collective gene pool. As Barton Childs, a pediatric geneticist at Johns Hopkins University, explained in a paper on genetic medicine written in the 1970s, "The secret of successful prevention is to know what, as well as how, to prevent. In addition the idea of the health of the species as a whole and of our genetic knowledge is one we should try to propagate."[93] Equally widespread during the 1980s were arguments about the cost-effectiveness associated with genetic testing and screening, which would not only eliminate supposedly deleterious genes in future generations but save millions of dollars by reducing rates of institutionalization and producing more industrious citizens with higher overall earning power. For example, two prominent geneticists touted the importance of calculating "the cost-benefit aspects of preventive programs," offering examples of the public costs saved by prenatal genetic screening. According to one of their estimates, "$83,000 in the case of males and $47,000 in the case of females" would be saved by preventing Tay-Sachs disease and Hunter syndrome through screening programs, and "$66,000 and $38,000 (assuming the total productive capacity of these infants would have been 20 per cent of the norm)" for males and females, respectively, in the cases of Down syndrome and trisomy 18.[94]

By the 1970s, the field had undergone a revolution in terms of gender composition, but its practitioners overwhelmingly were white and middle to upper middle class. According to the 2010 survey commissioned by the NSGC, women still account for 95 percent of the field, of whom the vast majority (92%) self-identify as white or Caucasian, 5 percent as Asian, and 1 percent as African American.[95] This profile has remained virtually unchanged since the 1970s and poses challenges for genetic counseling as it strives to engage a diverse multiracial and multiethnic clientele in twenty-first-century America.[96] As Nancy Callanan, acting NSGC president, exhorted in her 2005 presidential address, "There is an urgent need to increase not only the number of genetic counselors, but also diversity within the profession."[97]

Furthermore, debates continue about the virtues and weaknesses of nondirectiveness, which served as the banner ethos for genetic counselors intent on rejecting the paternalistic and prescriptive approaches of the preceding generation. Although nondirectiveness might have once been a useful guidepost, many assert that it is out of step with the complexity of their profession and hinders rather than helps. For instance, Jon Weil, who directed the genetic counseling program at the University of California at Berkeley from 1989 to 2001 and has been a leader in psychosocial counseling, contends that nondirectiveness "inhibits counselors from developing and using the full range of active counseling skills relevant to their goals for working with clients." Weil argues that nondirectiveness has been hard to shake off because it is "in the air, in the history" of genetic counseling, was a formative and accessible concept to many genetic counselors, and sustained the reticence among many to broach the issue of aborting affected pregnancies.[98] Resta, based at the Swedish Medical Center in Seattle, Washington, found that once he moved into counseling for cancer genetics, nondirectiveness became a barrier to helping clients who needed to learn about potential medical interventions following the results of tests for hereditary breast, ovarian, or colorectal cancer.[99]

As we move forward in the twenty-first century, discussions about the value of nondirectiveness are being overshadowed by the latest developments in genomic medicine. Genetic counselors are determining how to respond to the recent proliferation of direct-to-consumer (DTC) genetic testing, with many concerned about the psychological and ethical risks of purchasing personalized genetic information without any guarantee or interpretation or counseling.[100] Clients who pay out of pocket for services such as personalized genome mapping or genetic tests that are not reimbursable by insurance providers are likely to receive sensitive genetic information with little interpretation and no psychological counseling.[101]

Furthermore, ongoing discoveries about the role of genetics in common and chronic diseases—which can range from the new identification of single-gene disorders to calculated probabilities of genetic predispositions—have the potential both to heighten the demand for genetic counselors and to lessen the need for genetic specialists and subspecialists.

Genetic Risk

An Evolving Calculus

I n September 1977, Jack Murphy, a 22-year-old truck driver based near Flint, Michigan, wanted to get married. Jack's mother, maternal grandmother, two aunts, and one uncle all had suffered from Huntington's disease, a progressive illness characterized by movement disorders, cognitive problems, and psychiatric manifestations.[1] The most common inherited neurological disorder, HD is autosomal dominant, meaning that individuals need to receive only one copy of the HD gene from either parent to be affected. Before arranging the nuptials, Jack and his fiancée scheduled an appointment with the Heredity Clinic at the University of Michigan.[2]

Although Jack exhibited no symptoms suggestive of HD, he was worried about his family history and was especially troubled by the physical and mental unraveling of younger brother Roger. About one year earlier, Roger had noticed the "onset of a tic involving primarily the left eyelid, but also involving the right" and sought a consultation at the University of Michigan's Heredity Clinic, now called the Medical Genetics Clinic. Soon after, Roger sensed that his memory was failing, and he began to spontaneously drop glasses, fall off chairs, and succumb to "outbursts of temper" that escalated into beating his wife. On several occasions he felt so distraught that he tried to kill himself with an overdose of pills. By summer 1977, Roger was undergoing regular psychiatric counseling and was legally separated from his wife, who had sole custody of the couple's two children.

Jack and Roger knew that their mother's first clinical manifestations of HD appeared at the young age of 24, and they watched her spend the years preceding her death at age 45 severely debilitated and confined to a bed in a state hospital. For generations, and over many decades, multiple members of this extended family had developed HD. In fact, the brothers' grandmother was one of the first people with HD evaluated by the Heredity Clinic, in 1942; her family pedigree, which Heredity Clinic workers traced back to her great-grandmother, became part of an ambitious effort to locate all people with HD in Michigan's public hospitals and state schools.[3]

If Jack thought that genetic counseling would enable him to resolve the dilemma about whether to proceed with marriage and fatherhood, he was mistaken. Following a clinical evaluation and updating of his family pedigree, Jack opened a letter from the Heredity Clinic advising him that "there is no evidence that you are affected with Huntington's disease at this time. However, as we have discussed, the disease oftentimes does not present until after the age of 30 and possibly even after the age of 50, and it is impossible to tell at this time whether or not you will develop the disease. With your positive family history, the disease occurring in your mother, you are at a 50% risk of becoming afflicted with Huntington's disease." In other words, there was a strong probability that Jack would develop HD, but an equally strong probability that he would not. Neither did Roger, whose situation undoubtedly was direr, receive a clear-cut diagnosis. The medical geneticist who evaluated Roger noted, "One certainly has a feeling that the patient may be showing signs of early Huntington's Chorea, however, we feel the evidence is too soft at the current time to make a definitive diagnosis."[4]

Like all individuals with a biological family history of HD, the Murphy brothers lived at risk, in limbo, warily monitoring their own bodies and behaviors for any signs of lack of coordination, irritability, cognitive problems, or jerkiness. Formed the same year Jack and Roger consulted the Heredity Clinic, the federal Commission for the Control of Huntington's Disease captured the essence of being at risk in its multivolume report on HD in America: "Individuals at risk live in a state of uncertainty which imposes a heavy psychological burden. Not only must they bear witness to the painful decline of a parent; they must carry the burden of fear and anxiety that someday the same thing may happen to them."[5] In a similar vein, Nancy Wexler, a clinical psychologist with a family history of and vast professional expertise in HD, wrote in her now classic article "Genetic Russian Roulette" that genetic counselors should approach the "state of being at risk" as "qualitatively different from the state of knowing definitively either that one will be sick or healthy."[6] Without a test to determine whether they carried the HD gene, Jack, Roger, and thousands of other at-risk individuals lived in a liminal space often characterized by dread, denial, and chronic anxiety.[7]

Only six years after the Murphy brothers were seen at the Heredity Clinic, the dynamics of HD diagnosis changed dramatically. In 1983, James F. Gusella and colleagues located a polymorphic DNA marker for HD indicating the chromosomal location of the gene for the disorder on the short arm of chromosome 4 and which allowed the use of familial linkage studies for presymptomatic testing.[8] Ten years later, in tandem with the quickening pace of the Human Genome Project, HD researchers mapped the associated gene to the short arm of chromosome 4, making it possible to test individuals for the HD genetic mutation without the need for linkage studies. These researchers also demonstrated that clinical presentation and age of onset were associated with the number of times the three-nucleotide of CAG (cytosine-adenine-guanine) repeated beyond the normal range of 11–34 (now readjusted to 1–28). The Huntington's Disease Collaborative Research Group determined that as a general rule the higher the number of CAG repeats, the earlier the onset and more severe the clinical manifestations.[9]

These pathbreaking findings finally answered the vexing question about the etiology of HD, one of a small number of hereditary diseases caused by the alteration of a single gene. Yet they did not quell the anxieties of HD families. In its 1977 report, the Commission for the Control of Huntington's Disease had expressed great optimism about the impact of a presymptomatic test, which would "at least relieve the burden for half this group," freeing them to "be clumsy, to trip, to drop things, to get angry, to cry and feel sad—without the constant fear that these normal human experiences mark the onset of the disease."[10] Nevertheless, once this goal was attained, the reality of presymptomatic testing was much more nebulous.[11] Rather than neatly solving a genetic mystery, the new technology of presymptomatic HD testing created new quandaries: whether or not to be tested, how to react to a positive or negative diagnosis, and whether and how to involve or inform other family members in the testing process.[12] Indeed, scientists and commentators involved in and aware of dramatic advances related to HD have been surprised by studies in the United States and Europe indicating that only 20 percent of patients at risk for HD have chosen to be tested. In her moving memoir, *Mapping Fate: A Memoir of Family, Risk, and Genetic Research*, Alice Wexler (Nancy Wexler's sister) explains that she, like 80 percent of those sharing her at-risk status, "wasn't tempted, not seriously," by the test, preferring to live with the ambiguity of not knowing rather than the harsh certainty of a definitive result. For Wexler, this decision "meant learning once again to live at risk, with no thoughts now of a final release, no fantasies of freedom from the possibility of HD."[13] Presymptomatic HD testing also prompted questions about best practices for genetic counselors and the ethical dissemination of genetic information.[14]

The shifting risk calculus of HD as it crossed the transom of genetic testing and genetic counseling can shed light on broader changes in the meanings and applications of genetic risk in the twentieth century. Since the 1940s, calculating genetic risk has become increasingly targeted and precise, spurred on by expanding knowledge in and the specialization of medical genetics. Since the founding of the first heredity clinics in the mid-twentieth century, genetic risk assessment has changed markedly, enhanced by the information gleaned from observational studies of individuals with genetic conditions and by dramatic advances in biochemical genetics, cytogenetics, molecular genetics, computational biology, and proteomics. As genetics laboratories became more sophisticated, human geneticists designed novel technologies and instruments to assess genetic risk.[15] Nonetheless, developments in screening, diagnosis, and biostatistics have not easily correlated with how patients and clients evaluate their own risks.[16] More information about genetic conditions and greater precision in genetic risk, especially if few or no medical cures or therapies are available, do not necessarily result in less apprehension and in many cases can engender more.

For many patients and families who access genetic services—whether in a prenatal, pediatric, or adult setting—one of the end results is a diagnosable genetic condition.[17] For many more, however, the outcome is a numerical risk calculation or probability, determined using an assortment of technologies and tests based on family history and/or clinical symptoms or as part of prenatal and perinatal care. Genetic counselors are on the front lines of risk assessment and the delivery of this potent information to clients and families. Practitioners of genetic counseling have been instrumental in the formulation of equations and understandings of genetic risk. Especially in recent decades, they have challenged conventional thinking about risk behavior and produced resources for people like Jack and Roger, who find themselves in the unsettling state of being at risk.

Heredity Clinics and Their Clientele

Today genetic counselors can be found at most larger hospitals and many specialty clinics. In the 1940s, however, coming to a heredity clinic was a brand new experience. For the first time patients were seen by a human geneticist who tried to assess as accurately as possible the relationship between an actual or potential medical condition and genetic inheritance. Before the opening of heredity clinics, family pedigrees had been gathered in a less than systematic and medically oriented manner, sometimes by eugenic field workers intent on tracing degeneracy back through family generations, sometimes by a family doctor who intuited a hereditary pattern of disease, and sometimes by researchers studying a group of families and individuals affected by a specific condition such as HD, colorectal cancer, or polydactyly. During the early twentieth century, eugenic field workers gathered thousands of family pedigrees, mainly from ostensibly "unfit" individuals and their kin and with the explicit aim of justifying policies of institutional segregation or sterilization.[18]

The founding of the University of Michigan's Heredity Clinic elucidates how studying human heredity entered the world of academic science and medicine and introduced medical genetics to patients and physicians. The Heredity Clinic at the University of Michigan was established by Lee Raymond Dice, who had received his PhD in paleontology and zoology in 1915 after filing a dissertation on the ecologic dispersion of vertebrates in southeastern Washington, where he grew up.[19] After serving in World War I, Dice was hired as curator of animals at the University of Michigan's Museum of Zoology. During the 1920s, he studied the variation, behavior, and genetics of several species of the peromyscus (deer mouse). In 1934 Dice became the director of the Laboratory of Mammalian Genetics and devoted most of his energy to laboratory and field research.[20]

At some point in the 1930s, Dice, who was first exposed to eugenics during his undergraduate study with David Starr Jordan at Stanford University and his graduate work under Samuel J. Holmes at the University of California at Berkeley, decided that he wanted to extend his mammalian studies to humans. Initially interested in the hereditary aspects of epilepsy, he surmised that "the information about the heredity of convulsive behavior and other behavioral abnormalities in peromyscus that I and others secured by our studies in the laboratory of Vertebrate Biology suggested to me that some types of human epilepsy were perhaps also inherited in a fairly simply manner."[21] His curiosity piqued, Dice submitted a proposal to the Rackham School of Graduate Studies, which was favorably reviewed and ultimately led to the founding of a clinic for the "study of the heredity of human defects." Open to the public on November 12, 1941, the University of Michigan's Heredity Clinic straddled the basic sciences and the medical school.[22] The Heredity Clinic's institutional home was in the Laboratory of Vertebrate Biology, which pursued a spectrum of research from taxonomic surveys of local plant and animal species to laboratory studies of mice and monkeys.

Reflecting the low prestige of human genetics in biomedicine, the University of Michigan Medical School relegated the Heredity Clinic to a run-down two-story clapboard house that previously had served as interns' quarters.[23] Under Dice's direction, it launched clinically integrated studies of hereditary conditions including HD, neurofibromatosis, ectopia lentis, and Marfan syndrome. From the get-go, the small nucleus of faculty and staff at the clinic cultivated good working relations with physicians and researchers in the departments of internal medicine, neurology, ophthalmology, psychiatry, sociology, and anthropology. In 1946 James V. Neel became director, and 10 years later, as Dice readied for retirement, he completed the administrative transition of the heredity clinic to the medical school with the concomitant formation of the Department of Human Genetics.[24]

With the creation of heredity clinics under the aegis of academic and medical institutions, human geneticists began to streamline the process of patient intake and evaluation. In 1945, Dice oversaw the production of a manual that detailed procedures for genetic evaluation and included information about designated consultation rooms, payment policies, referral protocols, and requisitions. Dice fastidiously designed a system for the taking of pedigree charts, which involved moving from a draft form "in pencil on sulfite paper of canary color" to a standardized ink pen pedigree that began with the propositus—or the first person to be investigated in a family study—and then branched out to collect data on sibship A (the first set of siblings), indicated by an arrow, followed by sibships B, C, and any additional set of siblings. Each kindred or extended family unit was summarized on a file card, with pink for women and blue for men.[25] They were assigned a number and cross-filed chronologically and by reference to disease. To their credit, Dice and his colleagues sought to respect patient privacy: "At the first interview the propositus or his parent or guardian will be asked that the Heredity Clinic be permitted to interview other members of the kindred. If such permission is refused we usually cannot undertake a study of the kindred." All gathered information was to be treated as "strictly confidential."[26]

Although these methods were clinically driven and structured, their roots were directly traceable to the eugenics movement. Most notably, in the early twentieth century eugenicists in America and England codified and standardized the pedigree chart, developing the iconography of circles to represent females, squares to represent males, and letters to stand for specific conditions—such as A for alcoholism and F for feeblemindedness. In the 1940s, the pedigree chart migrated intact into medical genetics. As historian of genetic counseling Robert Resta has written, "The same tool—the pedigree—was used by eugenicists and human geneticists to demonstrate or prove their hereditarian claims."[27] Indeed, researchers at the University of Michigan's Heredity Clinic implemented the pedigree symbols devised by the Eugenics Record Office (ERO), based in Cold Spring Harbor, New York. From 1910 to 1939, the ERO was America's premier eugenics research organization, training the hundreds of eugenic field workers who traveled from state to state to produce pedigrees of supposedly "defective" families.[28] Utilizing a trait book developed by leading American eugenicist Charles B. Davenport, ERO workers assembled thousands of family histories and created an enormous database of inheritance information and patterns. Before coming to the University of Michigan, Neel visited Cold Spring Harbor and contemplated using the ERO's huge archive of family studies for a research project, but he judged the materials to be biased, badly organized, and of scant scientific value. For Neel, reviewing the ERO's stockpile of problematic data made "the magnitude of what had to be done before human genetics could be a respected discipline [become] brutally apparent."[29]

Michigan's Heredity Clinic—as well as the University of Minnesota's Dight Institute and Wake Forest University's medical genetics program—served three kinds of people: walk-ins, referrals, and human subjects. Walk-in clients were concerned about hereditary disease in their family, usually because they were planning to marry or have children soon. For the most part, these clients were characterized by "better than average income and education."[30] Sheldon Reed wrote that many clients had "learned about the Dight Institute through the newspapers or in some way or other, by talking with other people, so that we get a constant stream of people coming to the Dight Institute."[31] Some were curious about the risk of genetic defects in consanguineous unions and in

Sheldon Reed (*center*), his wife, Elizabeth Reed, and colleague David Merrell analyzing inheritance patterns of genetic conditions in 1949. A photograph of Charles Dight, the eccentric eugenicist who endowed the Dight Institute, looks down on them from the wall. *Source:* "What Do We Inherit? A Three-Way Attack on Heredity Problems," *Minnesotan* 3, no. 2 (1949): 3. Photo courtesy of the University of Minnesota.

situations of father-daughter and brother-sister incest.[32] Others fretted about the possibility of a child with developmental disabilities or a family history of albinism.[33] For example, one young man worried that his excessive masturbation would result in "Mongoloid" offspring.[34]

Geneticists also offered themselves as clients and test subjects. At the University of Michigan's Heredity Clinic, the entire staff and its family members participated in sessions to gather family pedigrees.[35] At the Dight Institute, Reed included himself and his wife in a set of illustrative cases. It turns out that the Reeds were Rh incompatible—Sheldon was Rh positive and his wife Elizabeth Rh negative. In the last weeks of pregnancy with her third child, Elizabeth had developed antibodies to her Rh-positive fetus. Concerned about developing erythroblastosis, Elizabeth opted for a premature cesarean section, after which she was surgically sterilized.[36] This decision was based on Sheldon's calculation of a 64 percent probability that their next child would also be Rh positive, a risk they judged as "too large to trifle with."[37]

The second group consisted of patients referred by physicians who wanted the expert input of a geneticist. These cases usually involved serious medical conditions with obvious and severe clinical manifestations and which were associated with a recognized genetic disorder. For example, of the 400 patients seen by the University of Michigan's Heredity Clinic during its first three years of operation, a majority were referred by three departments: ophthalmology (principally for retinitis pigmentosa and ectopia lentis), dermatology (principally for neurofibromatosis), and orthodontics (for various genetic abnormalities). By the mid-1940s, referrals were on the rise from the departments of neurology, psychiatry, pediatrics, and obstetrics and gynecology.[38]

Lastly, genetic counselors evaluated patients that can best be described as human subjects selected for studies that sought to identify the hereditary components of various conditions. Some of these studies verged on human experimentation and lacked the ethical protocols—such as informed consent—that researchers must adhere to today. In North Carolina, for example, Wake Forest University's medical genetics program dispatched researchers to the Smoky Mountains, where, often on foot with clipboard in hand, they gathered medical and personal information from members of communities ostensibly distinguished by high rates of "genetic defects" due to reproductive patterns of endogamous and isolated breeding. In Minnesota, the Dight Institute embarked on a large-scale study of the mental, physical, and psychological characteristics of the patients at the Faribault Home for the Feebleminded. A follow-up to the eugenic survey carried out in the 1910s, this family study generated the core of Reed's data on the status of people with mental retardation in mid-century America. During the same period, the Dight Institute facilitated a study of HD patients at Minnesota's Rochester State Hospital. Nevertheless, some of the early work performed by the heredity clinics can be considered precursors to contemporary and ethically acceptable studies of affected populations. For instance, one of the Dight Institute's early projects examined the heredity of breast cancer among female relatives of a set of consenting families, while the University of Michigan's Heredity Clinic pursued clinically circumscribed studies on HD, cystic fibrosis, sickle cell anemia, and thalassemia.[39]

Genealogies of Genetic Risk

The etymology of the word "risk" reaches back to the Latin term *riscum*. However, the term and concept of risk did not appear in earnest until the Middle Ages, associated with maritime ventures and the "perils that could compromise a voyage" on the sea. During this period, risk implied potential danger enacted by God or natural law without reference to the "idea of human fault and responsibility."[40] In the eighteenth and nineteenth centuries, with the growth of nation-states, theories of liberal individualism, and increasing belief in science and objectivity, the concept of risk shifted and secularized. It became embedded in the nascent sciences of probability and statistics, which sought to determine patterns, calculate norms, and ultimately provide predictive methods for human behavior and events.[41] In short, risk became a quantifiable construct, part of the calculus of an emerging management system organized to deal with the "hazards and insecurities induced and introduced by modernization itself."[42] At the same time, risk, which earlier could be associated with good outcomes—providence and luck—became defined almost wholly by negative connotations and outcomes. Over time, risk extricated itself from chance, which always retained an aura of unpredictability. Moreover, risk was becoming a biopolitical concept, concerned with how individuals managed their own bodies and how larger entities, such as states and societies, managed populations.[43]

As risk became central to understandings of modern life and behavior, experts and scholars from various disciplines began to articulate algorithms and frameworks to calculate, measure, and assess risk. The most well-known example of the development of expert risk knowledge is the rise of actuarial science associated with the insurance industry. After World War II, risk studies expanded, spurred on by concerns about the potentially negative impact of new technologies and products, ranging from nuclear weapons to environmental pollutants, on human beings and society. Notably, research on risk experienced a meteoritic boom from the 1960s to the 1980s, rising in frequency in scholarly journals between 1966 and 1982, especially after the early 1970s.[44] Drawing from mathematics, psychology, and sociology, a cross section of scholars probed the relationship between quantitative or numerical risk values and qualitative or social risk perception. This research, broadly grouped under risk analysis, explored various dimensions of environmental, occupational, technological, and related health risks.[45] The paramount assumption in this literature was that individuals and societies behaved as rational decision makers, operating in accordance with utilitarian principles of self- or group maximization.[46] In this framework, actors determine what level of risk is acceptable and make decisions accordingly about a particular course of action. In keeping with the historical context in which he practiced, Sheldon Reed embraced this approach, believing that his clients rationally evaluated their genetic risk and almost always acted in a manner beneficial for their families and society at large. As he explained, "The idea of a calculated risk makes sense to most everybody."[47] Risk scholars Christina Palmer and François Sainfort have identified this orientation, which dominated until the late 1970s, as a cornerstone of the era of the "objective characterization of the magnitude of the recurrence risk."[48]

Over time, however, researchers interested in how people, especially patients in health care settings, evaluated risk information for decision making came to realize that the relationship between the quantitative and qualitative dimensions of risk was far from commensurate—the two seldom corresponded. By the 1980s, converging currents were challenging the commensurability model, particularly in the realm of genetic counseling. Researchers were demonstrating wide variations in risk perception and showing that patients rarely thought "in terms of the data-base and computational formulae used by genetic risk analysts."[49] Instead, people evaluated the same level of risk, even the exact same numerical representation of risk, in divergent ways whether or not they shared many demographic and social characteristics. As a British genetic counselor succinctly noted, "A specific, genetically determined disease can mean different things to different people."[50] F. Clarke Fraser, a foremost medical geneticist at McGill University, explained that "it is important to remember that a risk which seems reassuring to one person may be formidable to another and that a defect that seems trivial to one person may appear disastrous to another."[51]

Perhaps most influential, in 1979 Fraser, working with the epidemiologist Abby Lippman-Hand, published a series of landmark articles on risk, perception, and choice.[52] Based on analysis of qualitative interviews with clients undergoing genetic counseling, Lippman-Hand and Fraser debunked the assumption that genetic counselors and clients share the same understanding of genetic risk values and that "these rates are provided in ways that are precise and, in themselves, useful as a basis for action." They wrote that "ambiguity arises because there is no necessary one-to-one correspondence between a diagnosis and its implication. Because of variable expressivity, a range of possible 'burdens' may ensue from one specific condition."[53] They also found that clients tend toward simplification and think about risk in binary terms. Jon Weil, who ran the University of California at Berkeley's genetic counseling program for many years, has explained that "binarization involves a cognitive simplification in response to being at risk, with the emotional burden and responsibility for critical decisions this involves."[54] In response to their findings, Lippman-Hand and Fraser concluded that genetic counselors needed to develop a more nuanced model of genetic risk assessment, one that underlined "the heuristic processing of complex information rather than one that assume[d] a traditional cost-benefit (risk-burden) approach."[55] More recently, Palmer and Sainfort have extended this analysis, arguing that clients and patients tend to evaluate risk along a presence-absence model, not according to whatever numerical odds ratio or percentage is provided in a genetic counseling session. They assert that "the *presence* of both uncertainty and adversity is the relevant and necessary aspect of our view of risk, not the *amount*."[56]

Taking cues from this kind of research, in the past two decades scholars from many disciplinary backgrounds have challenged the rational choice model that held sway for much of the twentieth century.[57] There is a growing body of scholarship that examines the emotive dimensions of risk and proposes that risk—even the prospect of taking a risk—should be understood as a feeling replete with enigmatic psychological and cognitive meanings.[58] From another angle, social theorists Carlos Novas and Nicholas Rose situate the consolidation of the concept of genetic risk in the 1970s, tied to the rise of prenatal genetic diagnosis and the construction of the at-risk individual, which they also refer to as the "somatic individual."[59] Read alongside pioneering work in anthropology on the relationship of culture and risk, these more multifaceted understandings of risk and its meanings are particularly relevant to genetic counseling.[60]

One clear trend over the past 60 years has been the gradual lowering of the thresholds of low, intermediate, and high risk. Whereas Reed and his contemporaries evaluated 10 percent risk of a genetic disorder as low, by the 1970s 10 percent had become high for many. In her research on the orientations of genetic counselors for *The Tentative Pregnancy*, published in 1986, Barbara Katz Rothman found that "a risk rate of one-in-fifty is called high or very high by almost half of the counselors I interviewed—and by only 20 percent of the earlier group of counselors."[61] The creeping tendency has been to treat any level of risk as evidence of risk, as opposed to its absence, even if that risk is calculated to be 1 in 400, or less than 1 percent. Discussing her role as a psychologist and genetic counselor at Columbia Presbyterian Hospital in New York, Diana

Puñales-Morejon has observed that some patients, especially wealthier patients, want a guarantee of certainty when it comes to personal and reproductive health. Puñales-Morejon, who graduated from Sarah Lawrence in 1985 and has since served as a classroom and clinical instructor for the program, reports that for many, "no risk is tolerable."[62]

The changing calculus of risk assessment can be traced through the evolving practice of genetic counseling and the experiences of counselees over the past seven decades. Initially, genetic counselors relied on the foundational tool of Mendelian risk assessment, which calculates the probability, based on as extensive and accurate a family pedigree as possible, that a person will develop an inherited condition. When Dice, Reed, and their contemporaries founded heredity clinics in the 1940s, Mendelian risk assessment was the fundamental tool at their disposal. Mendelian risk focuses on autosomal dominant disorders such as HD, recessive conditions such as Tay-Sachs disease and cystic fibrosis, and X-linked disorders such as hemophilia and Duchenne muscular dystrophy. Writing about the early years of genetic counseling, Reed described this risk analysis as "the old fashioned type where all one could do was to give a risk figure for the repetition of the particular defect in subsequent offspring."[63]

If these patterns were seductively neat, genetic counselors realized that many conditions did not conform to stock Mendelian ratios of 1 in 2 or 1 in 4. In particular, Down syndrome, which until the 1960s was alternately called mongolism, Mongolian idiocy, or mental retardation, stubbornly refused to fit into these probabilities. When heredity clinics opened their doors in the 1940s, genetic counselors regularly met with parents who had one child with Down syndrome and were eager to know the likelihood that their next child would be similarly affected.

In 1933, based on a statistical study of 150 sibships, British geneticist Lionel S. Penrose suggested that there was a significant association between the advanced maternal age of mothers and higher birth rates of "mongolian imbeciles."[64] Reed wanted to better understand this likely correlation in terms that he could present to his clients. This puzzle led him and his Swedish collaborator, Jan Böök, to formulate the concept of "empiric risk," which relies not on theories of Mendelian inheritance but on the observation of patterns of disease experience among families and larger groups. Reed and Böök debuted this term in the *Journal of the American Medical Association* in 1950, and it remains central to genetic counseling today.[65] Drawing from Böök's data on mongolism in a North

Swedish population, they calculated an overall frequency of .09 percent among women in the 20–29 age bracket but a much higher "statistical chance of about 1 to 6 percent" in women over 40. They concluded that "there remains no doubt that maternal age is an important factor in the etiology of Mongolism."[66] Several more Scandinavian researchers employed the concept of empiric risk to calculate the likelihood of inheriting low-grade mental deficiency and anencephaly in bounded population groups.[67]

The concept of empiric risk also resonated with Reed's study of the genetics of breast cancer. [68] Undertaken by the Dight Institute in 1943 and funded by the American Cancer Society, this study sought to determine the potential hereditary component of breast cancer among 173 women with reported family histories.[69] After taking over the institute, Reed continued this longitudinal project, releasing his findings in 1950. He asserted that, barring the eye cancer retinoblastoma and the neurological and dermatological cancer neurofibromatosis, there was not the "slightest relationship between the concentration of cancer in families and the true Mendelian ratios found in albinism, hemophilia, and other hereditary conditions."[70] Instead, he suggested that individuals could "inherit a susceptibility to some kinds of cancer." One of Reed's major conclusions was that there were "four times more deaths from breast cancer among sisters of breast cancer patients than is believed to exist among women of the same age group in the general population."[71]

The search for the hereditary influences in breast cancer and the condition then called mongolism underscored the strengths and limits of empiric risk calculations. Reed and Böök were careful to clarify that empiric risk was observational, was based on epidemiological patterns of clinical presentation in specific populations, and did "not imply any particular theory about the etiology."[72] In the late 1950s, the University of Michigan's Neel expanded the literature on empiric risk analysis. He offered the example of a researcher who had completed family histories of 100 individuals affected by harelip, all of which had normal parents, and a combined total of 300 siblings, 15 of whom exhibited harelip. He calculated that the "the empiric risk of encountering this trait among the siblings of an affected individual" was "5 per cent [or 15/300], and this risk could then be applied to the probability of any future sibling of such an individual exhibiting the defect." Neel also stressed that empiric risk was silent on etiology; it was based on observation of expressed traits and not on the potential multifactorial genetic causes of a disorder. Summing up, Neel wrote that "empiric risk figures are essentially pragmatic probability statements based on accumulated medical statistics."[73] Geneticist Kenneth K. Kidd compared Mendelian and empiric risk models in a series of articles in 1979, concluding that "empiric recurrence risks are considered and used for those familial traits and disorders which have an unknown (or at least unclear) etiology and do not show simple mendelian patterns of inheritance."[74]

In 1959 Jérôme Lejeune identified the relationship between trisomy 21 and Down syndrome. This revelation was quickly followed by discoveries of the function of chromosomal anomalies in a range of conditions, including other trisomies and sex chromosome disorders such as Turner and Klinefelter syndromes. Even as Lejeune showed that Down syndrome (95% of the time caused by nondisjunctional trisomy 21) resulted from a chromosomal anomaly and was not an inherited disorder, the concept of empiric risk remained a staple of genetic risk assessment. Specifically, the mathematical work attached to empiric risk provided a model for the compilation of index cases of an inherited condition that could populate a reference database. In coming decades, this same logic was applied to conditions with Mendelian inheritance patterns, such as Tay-Sachs disease, in order to assemble data from a large set of affected populations, including those defined primarily by ethnicity and race.

In addition to Mendelian risk and empiric risk, genetic counselors also relied on techniques of Bayesian risk assessment, which allow for the incorporation of information from different sources (age of onset, enzyme levels, number of healthy children, etc.) to determine a person's genetic risk factor. Weil has emphasized the significance of Bayesian analysis, which serves as a "standard method for evaluating the risk of having a child affected by a recessive or X-linked disorder or a dominant disorder with incomplete or age-dependent penetrance even in small, otherwise simple families."[75] Updated as new data are generated, Bayesian models are particularly well suited for large-scale non-Mendelian genetic risk analysis, understood as tentative and changing as kindreds are expanded and altered.[76]

The three models of risk assessment—Mendelian, empiric, and Bayesian—developed concurrently with evolving genetic technologies and tests, so that by the 1970s, risk prediction had changed markedly from the early years of heredity clinics. In particular, the advent of amniocentesis, which allowed for the diagnosis of fetal chromosomal anomalies, and the launching of screening programs for diseases such as Tay-Sachs disease and sickle cell anemia helped to transform genetic risk from an abstract estimation to a more precise calculation. As Reed told a group of physicians during this period of change, genetic counseling was no longer "a simple guessing game in which the counselor interprets the genetic or environmental risks to the couple who have had an abnormal child," but "a preventive medicine situation of the most critical importance for the family."[77]

What's My Risk?

When American housewives opened up their copy of the popular woman's magazine *McCall's* in December 1948, they were greeted by a smiling mother, cherubic child, and stick figures attached to pedigree charts. These images accompanied the article "Whose Little Girl Are You?," which surveyed the expanding knowledge of human heredity and its impact on American families. Highlighting the University of Michigan's Heredity Clinic, this article outlined the basic rules of Mendelian inheritance as illustrated by the recessive condition of albinism, the dominant disorder of HD, and patterns of eye color. Readers learned that "your own inherited traits and characteristics—and those you will bestow on your children are no longer a matter of guesswork. In most instances, they can now be scientifically and mathematically calculated."[78]

In newspapers, in magazines, and on the radio, early genetic counselors, such as Dice and Reed, introduced concepts of genetic risk and genetic counseling to tens of thousands of Americans. Clearly, the message resonated. According to Neel, who had just joined the Heredity Clinic, this *McCall's* article "precipitated a flood of requests from all over the nation for advice on personal problems in human heredity." In the weeks following its publication, "two to five letters per day have been received requesting advice on specific problems of family heredity. This illustrates the need which people have for advice about their heredity."[79]

While this article enhanced the profile of the University of Michigan's Heredity Clinic, in another part of the upper Midwest the Dight Institute was garnering attention in newspapers and radio broadcasts. Reed frequently was quoted as the resident expert on all matters relating to genetics. For example, a 1958 article attested to his mastery of genetic risk: "In most rare afflictions Dr. Reed is able to state with precise mathematics what chance (one in four or one in two, for example) this particular couple has of producing a child so afflicted. Each birth faces the same

22

Whose little girl are you?

Here are the known facts about heredity.
They reveal what you can—and cannot—
inherit and pass along to your children

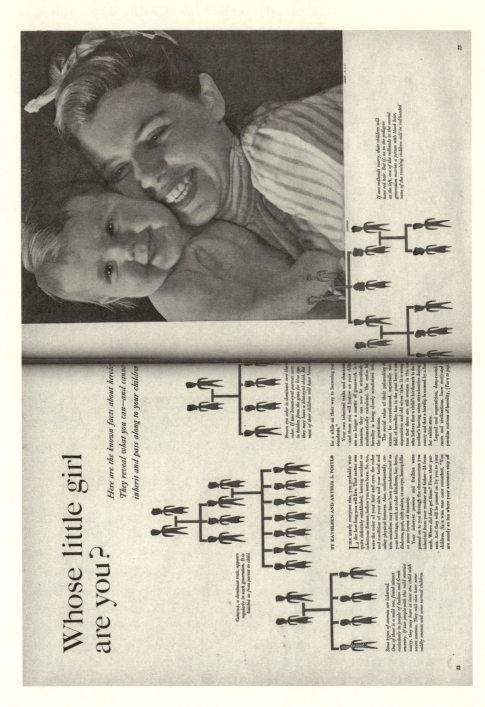

Cataract, a dominant trait, appears regularly in each generation. It is handed on from parent to child

Some types of anemia are inherited. One of these is a mild one, found almost exclusively in people of Italian and Greek ancestry. If two people with this mild anemia marry, they may have all their children severe anemia. They will also have some mildly anemic and some normal children

Brown eye color is dominant over the color. If two brown-eyed parents carry in hidden form the gene for blue eyes, they may have a blue-eyed child. But most of their children will have brown

BY KATHLEEN AND ARTHUR S. POSTLE

LIKE nearly everyone else, you probably wonder quite definitely how you will live. That matter was not infectious disease, before you were born. So, too, were the color of your hair and eyes, the color and condition of your skin and your facial and other physical features. Also, unfortunately, certain maladies may have been predetermined as your heritage, such as color blindness, hay fever, diabetes, gout, cleft palate, cross-eye, hemophilia or some forms of insanity.

Your inherent powers and frailties were passed on to you through the chromosomes you inherited from your mother and father—24 from each. Where did they get them? From their parents. And they will be passed on by you to your children. As a wise man once remarked, "You are merely in tan where your ancestors stop off

for a while on their way to becoming your descendants."

Your own inherited traits and characteristics —and those you will bestow on your children —are no longer a matter of guesswork. It is mathematically calculated. The entire subject of heredity is being closely scrutinized today by the geneticists.

The social value of their painstaking research cannot be overestimated, especially since the field of heredity has in the past been shrouded in superstition and old wives' tales. It is strange but true that there are still women in the country who believe that a child's birthmark is due to the mother's having eaten strawberries during pregnancy, or that a rabbit is caused by a fear of a rabbit stew.

Legend and superstition, deep-rooted in races and civilizations, have motivated many peculiar notions of heredity. [Turn to page

If two redheads marry, their children will have red hair. But if, as in the pedigree at the left, one of the redheads in the second generation marries a person with black hair, none of the resulting children will be red-haired

odds and results of previous births have no effect."[80] Reed also provided a steady stream of snippets of genetic information for the "Mr. Fixit" self-help columns in the *Minneapolis Tribune*, which featured apprehensive questions from readers about the role of heredity in hair and eye color of potential children and the offspring of incestuous affairs.[81]

Three years earlier, Reed's *Counseling in Medical Genetics* had hit the shelves.[82] The first popular educational manual on genetic counseling, this book sought to teach readers about the philosophy of and reasons for genetic counseling. It covered more than 20 topics—including mongolism, blood genetics, allergies, and the genetic effects of radiation—and presented the genetic counselor as a professional poised to provide scientific information and compassion to clients concerned about genetic problems. One of the overriding messages of this book was that genetic risk was universal: "Almost every family has some troublesome situation directly related to the heredity of one or more of its members."[83] As a columnist for the *Minneapolis Tribune* reminded readers, "All of us carry some recessive genes."[84] Showering medical and lay publications with ads for *Counseling in Medical Genetics*, the publisher W. B. Saunders sought to appeal to Americans far and wide: "In the largest cities—in the smallest country towns—anxious parents ask the doctor, 'What part does heredity play? What are the chances of our child being abnormal?'"[85]

Each year the Dight Institute and the Heredity Clinic advised a growing number of clients who were eager to understand "the size of the risk."[86] For example, in 1957, a mother from Brooklyn, New York, wrote to Reed. Her stillborn infant's appearance was consistent with mental retardation, and she wanted to know the likelihood that her next pregnancy would be similarly affected. In her response letter, she thanked Reed but also expressed her resignation to the risk figure he had calculated: "I was somewhat encouraged by what you said, even though 'somewhat less than 1 in 7' doesn't make me nearly as happy as 1 in 1,000,000 would."[87]

While hundreds of Americans sought out genetic risk information in

Opposite: This article, published in the women's magazine *McCall's* in 1948, mentioned the heredity counseling being offered at the University of Michigan's Heredity Clinic and, according to James Neel, "precipitated a flood of requests from all over the nation for advice on personal problems in human heredity." In the 1940s, images and explanations of family inheritance patterns and human genetics began to enter popular culture through magazines, television, and radio. *Source:* "Whose Little Girl Are You?," *McCall's*, December 1948, 22–23.

person or through correspondence, others evaded the attention of genetic counselors. Particularly when a genetic disease had caused great pain and distress in an extended family, some members were reticent to divulge anything. This was a fairly common response among HD family members, who had endured more than their fair share of stigma.[88] For example, researchers struggled to extract information from Jack and Roger Murphy's maternal great-uncle, one of the first members of their kindred contacted by the University of Michigan's Heredity Clinic in 1944. The clinic's secretary wrote imploringly in a letter, "We have tried several times to get in touch with you but have not had much success. We are very anxious to have an opportunity to examine your children. It is hoped that the examination, which we will provide at our own expense, will give the necessary information concerning the likelihood of these children to inherit their mother's defect."[89] In the case of another HD kindred, the Heredity Clinic staff penned a letter to one family member in the hopes of reaching her sister, who replied, "I can't tell you, how very sorry I am, Sister N hasn't written you. She seemed very skeptical at first, and didn't see why she should give out family information. After explaining you were doing research for the medical profession and would be strictly confidential she seemed more friendly and willing to cooperate."[90] To circumvent these kinds of impasses, Reed's staff often contacted non-affected family members first. As he told a group gathered at the Dight Institute for a lecture in the 1960s, "We find that in doing research on deleterious traits you get pretty good information from the in-laws about the affected members of the family, but the blood relatives are not quite so free with the information. Now why is that? It's obviously due to the fact that those who are likely to get the trait deplore the situation and are not telling."[91]

Reed complained that many clients who came to the institute eager for genetic risk information only wanted to hear about pleasing topics like eye color and European ancestry and were less disposed to acknowledge an actual risk of genetic disease. He described this incongruity: "Everybody has some kind of colored eyes and generally most people are fairly satisfied with their eye color, whereas they're not so interested in acknowledging the fact that they might have inherited the gene for Huntington's chorea, for instance. . . . We find that clients at the Dight Institute do not really wish to learn that their problem has a hereditary or genetic basis. For instance, they want to be told that their albino child is just an exaggerated Scandinavian."[92] While clients and patients varied in their willingness to share information with a genetic counselor, they also had markedly different reactions to genetic risk figures. They received genetic risk as a percentage (e.g., 50%), proportion (e.g., .5), chance (e.g., 1 in 2), or odds ratio (e.g., 1 to 1) and then weighed that number in accordance with a complex array of medical, familial, social, and personal criteria.[93] Couples who already had a child affected by a hereditary disorder evaluated risk figures differently than those who were childless. Reed noted that if the couple had not yet produced a "normal" child, they usually would keep trying.

Acceptable levels of risk varied from client to client. For some, there was no difference between 1 and 25 percent: "they don't oftentimes distinguish between one chance in four and one chance in one hundred. It's all the same to them."[94] Reed's general rule was to present 3 to 5 percent as a small risk, too negligible to alter any reproductive plans.[95] Generating the most equivocation was the risk figure of 25 percent or 1 in 4, which Reed viewed as a borderline figure: "it is clear that a 25 per cent risk of a repetition of a serious defect is about the threshold point for most families. If the risk is greater than 25 per cent, they are likely to decrease their reproduction significantly; whereas, if the risk is below 25 per cent, they are quite likely to ignore the risk."[96] Almost all couples interpreted 50 percent as too high in emotional and financial terms. Yet in a few cases, a small risk was sufficient to be used as a rationale for couples to refrain from having children. In 1967, Lee Schacht, the genetic counselor for the Minnesota State Board of Health's human genetics unit (which maintained close links to the Dight Institute), told a *Wall Street Journal* reporter, "You'd be surprised how many couples use this as an excuse for not having any more children," adding, "I'll quote them a risk of 2% or 3%, which isn't very high, and they'll push me to raise it."[97]

During the 1940s and 1950s, genetics clinics could only calculate risk based on family pedigrees and clinical observation. By the 1960s, refined biochemical techniques to test for certain enzyme and metabolic disorders, such as PKU or maple syrup urine disease (MSUD), and the advent of cytogenetics and chromosome analysis ushered in significant changes in genetic diagnosis and genetic risk analysis.[98] Karyotyping—which initially involved cutting, pasting, and ordering microphotographic images of chromosomes on a sheet of paper—provided a powerful visual diagnosis. Now clients could see the molecular manifestation of Down or Klinefelter syndrome. Once these technologies were incorporated into amniocentesis in the 1970s, prenatal diagnosis created a new landscape

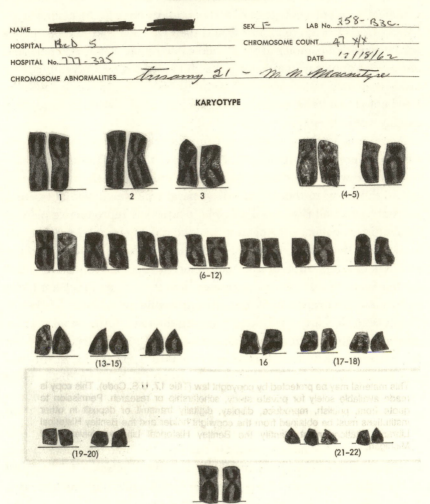

CYTOGENETICS LABORATORY

CHROMOSOME ANALYSIS

NAME ███████████ , ██████████   SEX _F_   LAB No. _258- B3c._

HOSPITAL _BeD 5_   CHROMOSOME COUNT _47 XX_

HOSPITAL No. _777-325_   DATE _12/18/62_

CHROMOSOME ABNORMALITIES _trisomy 21 ~ M. M. Macintyre_

KARYOTYPE

1   2   3   (4–5)

(6–12)

(13–15)   16   (17–18)

(19–20)   (21–22)

SEX CHROMOSOMES

This 1962 chromosome analysis, performed at (Case) Western Reserve University's Cytogenetics Laboratory, used cut-and-paste karyotyping to diagnose trisomy 21 (Down syndrome) in a patient. The University of Michigan's Heredity Clinic was an early adapter of chromosome analysis, and the addition of karyotype analysis to clinical genetics created a new iconography for chromosomal and sex-linked disorders. *Source*: Box 32, Kindred 8436, Adult Medical Genetics Clinic Records, 4446 Bimu 2, Bentley Historical Library, University of Michigan. Access approved by University of Michigan Institutional Review Board, HUM00012519.

for genetic risk assessment and reproductive decision making. Reed aptly captured this shift: "In the past, the human geneticist or physician could only give the 'probable risk' of a birth defect. Now with in utero detection, the parents can be informed that they will have either an effected or a normal child. The risks are no longer one in four or one in three, but are 100 percent or 0."[99] Indubitably, these momentous developments played a crucial role in the lowering of thresholds of acceptable risk noted by Katz Rothman and facilitated the binarization and simplification of risk values discussed by Lippman-Hand, Fraser, and Weil.

Even if clients received more accurate risk calculations, this did not necessarily mean they would interpret the odds and percentages presented to them in a predictable manner. Over time, genetic health practitioners have learned that people weigh the same percentage or category of risk very differently based on their cultural, religious, moral, and individual values. Moreover, individuals with exceedingly similar backgrounds—including identical twins—can respond quite differently to genetic risk calculations. Not surprisingly, conversations about genetic risk often can be very emotionally charged because they involve and invoke attachments to living and deceased family members and are entangled with expectations about one's own life and future. Compounding this sensitivity are the negative stereotypes attached to many genetic disorders and the legacies of secrecy and silence experienced by many families with histories of hereditary conditions, such as the Murphy brothers and other HD at-risk clients.

Several studies have suggested that genetic counselors tend to emphasize genetic risk and elevate its importance.[100] Given that genetic counselors appreciate the subtleties of risk calculation, this could be interpreted as professional duty. But once this tendency intersects with the propensity of clients to evaluate risk—whatever the figure—as binary, and even infinitesimal risks as unacceptable, obvious miscues and crossed communication can occur.[101] For many genetic counselors this means that they have two sometimes incongruous goals during a consultation. On the one hand, they need to present risk information as accurately as possible, and on the other, they need to provide a sounding board for clients to evaluate risk calculations and probabilities according to their own unique ensemble of values.

Families at Risk

In genetic counseling, risk is almost always a family affair: "The diagnosis of genetic risks and constitutions is never an individual matter. It is always in principle something that affects a wider network of people beyond the one individual who may have been tested."[102] This was certainly the case with the Murphy brothers. In August 1985, their maternal cousin, Kimberly Thompson, called the University of Michigan's Heredity Clinic, renamed the Adult Medical Genetics Clinic, to speak with a genetic counselor.[103] Just 25 years old and aware of her family members who had been diagnosed with HD at the University of Michigan, including her mother, three aunts, and several cousins, Kimberly said that she was experiencing "trouble with her at-risk status."[104] A resident of southeastern Michigan who had taken the lead in coordinating a local chapter of an at-risk group for HD, Kimberly also was frustrated that her sister Marsha was exhibiting symptoms suggestive of HD but refused to see a doctor. The consultation notes suggest that the genetic counselor offered guidance consonant with Kimberly's situation and long family history of HD.[105]

The counselor who answered Kimberley's call in 1985 would have been one of a handful of master's-degreed counselors, trained at Michigan and elsewhere, who were familiar with support groups for at-risk HD family members. Catalyzed by the activities of the Committee to Combat Huntington's Disease and the federal Commission for the Control of Huntington's Disease, emergent support networks owed a great deal to the tireless activism of Marjorie Guthrie, the wife of American musical icon Woody Guthrie.[106] Marjorie watched her talented and creative husband endure the psychological and physical ravages of HD, and after his death in 1967 she devoted her energies to creating support services for families and raising awareness among scientists and mental health advocates about the disease.[107] On the local level, the establishment of support groups often resulted from the pioneering work of a small but dynamic group of genetic counselors, such as Dorene Markel, who had found jobs in their chosen field and relished working closely with patients and families affected by genetic diseases and their at-risk status.

Markel graduated from the University of Michigan's master's degree genetic counseling program in 1983, a member of its third cohort. The creation of Michigan's program was due largely to the efforts of Diane

Baker, a Sarah Lawrence alumna, who had been hired by the Michigan Community Department of Health in 1979 to provide counseling in genetics clinics around the state. Along with Traverse City and Marquette, one of Baker's sites was Ann Arbor, more specifically the University of Michigan pediatrics department, where her services were quickly recognized and appreciated. Soon after Baker's arrival, Robert Erickson, a pediatric geneticist, asked her about the possibility of starting a genetic counseling program at Michigan. Baker responded affirmatively and looked to the Sarah Lawrence and the University of California at Berkeley programs as models. With Baker's groundwork and a concerted push by key physicians and professors in the Medical School, in 1980 Michigan accepted its inaugural class of genetic counseling master's students.[108] Baker eventually became the director of Michigan's program, nurturing its successful growth until 2001, when she moved to Washington, D.C., and passed the mantle to Beverly Yashar.

Markel was the first person hired with the title of genetic counselor by the Medical School and was instrumental in establishing HD support groups in southeastern Michigan. When she started in her position, Markel was shocked to find a glaring lack of materials related to the psychosocial component of HD, especially the limbo status of being at-risk. In response, and of the belief that patients would benefit from speaking to other at-risk individuals outside their own families, Markel created a therapy group for HD patients. She also worked to set up a presymptomatic screening program. Her efforts enabled families to navigate risk in a safer emotional environment that was being purged of the shame and isolation so commonly attached to HD from the 1930s to the 1970s.[109] Markel continued to apply her skills in the 1990s, when she participated in the delivery of the country's earliest set of hereditary breast cancer or BRCA1 test results to patients at the University of Michigan hospital. Aware of the sensitivity of the risk information contained in the test results, Markel clearly remembers the first patient she counseled, a woman whose family history and pedigree chart had suggested a high probability of developing breast cancer. Markel and her team consulted with bioethicists and other experts around the country before informing the patient that her BRCA1 test results actually revealed that her likelihood of developing breast cancer was very low.[110] Although this outcome might have offered relief to the patient, who cancelled a scheduled bilateral prophylactic mastectomy based on her revised risk status, it did not resolve the broader issue of genetic risk for her extended female kin. Furthermore, as genetic counselors have shown in many studies, negative genetic test results for BRCA1 and BRCA2 have the potential to produce a sense of survivor's guilt among family members unaffected by the mutation who then ask, why my sister or aunt and not me?[111]

CHAPTER THREE

Race

Tense and Troubled Relations

On February 11, 1951, the Sunday service of the Cathedral Church of St. Mark in Minneapolis featured an address by one of the country's preeminent human geneticists, Sheldon Reed. Director of the University of Minnesota's Dight Institute, Reed spoke on a topic near and dear to his heart: the destructive force of biological racism and the vital need for racial harmony in American society. Reed's special lecture to more than 600 parishioners at this interracial service, which was sponsored by the Minneapolis Urban League and the Minneapolis Interdenominational Ministerial Alliance, among many other local civic and religious organizations, was a stark condemnation of white discrimination against black Americans.[1] In "All Men Are Brothers under the Skin," Reed asserted that any scientist or layperson who employed biology to justify ideas of racial superiority and inferiority was a misinformed bigot. Human genetics had demonstrated definitively that differences in skin color were the simple result of physiological responses to patterns of migration and settlement across the continents over the past millennia. He asked rhetorically, "Are there important differences between the blood of Negroes and whites?" and countered emphatically, "The answer is no!" Reed concluded that the "eventual complete amalgamation of all the races now in this country must and will come about. Let us all strive to eliminate racial prejudices and conflict during the process of amalgamation. Let us scramble our American eggs in peace!"[2]

Yet, only one month before this impassioned talk Reed conveyed a very different message when he evaluated the racial traits of a 2-month-old biracial girl being considered for adoption. The inquiry about this infant's racial classification was one of hundreds presented to the institute by potential adoptive parents, birth families, and welfare agencies from the 1940s to the 1970s. In this instance, a social worker from the Hennepin County Welfare agency called Reed about a "striking looking

Irish girl" who produced a child with a "light mulatto" man who happened to be a well-known local jazz musician. Willing to tolerate the stamp of illegitimacy, the birth mother wanted to "keep the baby," to which "she was strongly attached." In due course, the birth mother's sister-in-law brought the child to Reed to determine if "it wouldn't be too dark" for the family to keep. During two consultations, he noted that this infant girl with blue eyes, "straight, fine, hair," and "no pigment genes" nonetheless exhibited African ancestry because of her skin's "slight olive color" and "slight color on the last phalangeal joint of the fingers." Despite Reed's proclamations about racial sameness at the Cathedral Church of St. Mark, when practicing genetic counseling at the Dight Institute he adhered to the one drop rule of racial hypodescent. Ultimately, Reed judged that the newborn was indeed "too dark" to remain with her birth mother, who reluctantly relinquished the baby girl for adoption through an Illinois agency, an action endorsed by Reed.[3]

The tension between Reed's antiracist sermon and his racialized adoption assessment sat at the core of mid-twentieth-century genetic counseling and medical genetics. Like most of his colleagues, Reed averred that the vast majority of phenotypic or expressed traits were due to multifactorial inheritance and that disparities between blacks and whites in terms of income, education, and housing stemmed from discriminatory social policies. As Reed intoned to listeners of the twin cities radio station WDGY in "A Scientist Looks at the Races of Man," America's "racial problems are due to faulty education and not to the unimportant biological differences between the races."[4] Reed went so far as to claim that he would be happy if one of his children were to marry interracially and produce mixed-race children. In a speech Reed gave in 1954, he posed the question "Would you want your daughter to marry a Negro?" and boldly replied,

> I have a daughter and she is the apple of my eye! My answer is that
> I want my daughter to marry whom she chooses. If her choice be a
> Negro, he has my approval in advance, as I trust her judgment com-
> pletely. Presumably her choice would be intelligent, of good character
> and with comparable education. It would be more important that their
> thoughts and actions be compatible, than that their skin color and hair
> shape be identical. If her choice should be a Negro, there would be no
> grounds for fear of biological disaster, and it is to be hoped that they
> would not suffer from the social bigotry of others.[5]

Reed joined a prominent group of scientists, including anthropologist Ashley Montagu and geneticist Theodosius Dobzhanksy, when he railed against racism and doctrines of racial purity.[6] After World War II, scientists across the world discarded discrete racial categories and, even more unsparingly, racial hierarchies. Yet this rejection left a void when it came to human differentiation and classification. To continue their research, social and human scientists, especially geneticists, needed a taxonomic system to organize human beings in terms of time, space, and biology.[7] More often than not they turned to the concept of population, which allowed for the mapping of human differences on a spectrum of overlapping gradations linked to ancestry, geography, and migration. But, as science studies scholars have shown, population was never a neutral biogeographic term.[8] Instead, it carried a series of embedded and pliable racial precepts into human genetics, with consequences that shaped and continue to shape the assumptions and contours of genetic counseling, genetic screening, and pharmacogenetics.[9] Framed by the population construct, from the 1940s to the 1960s early genetic counselors participated in a complicated dance between race and population, sometimes advocating for racial and interracial associations, other times reinforcing and reinscribing racial and ethnic differences.

The Race in Population

Human genetics has never been able to escape theories of race, even when seeking to transcend rigid categories of race and identity.[10] The eugenicists of the early twentieth century promoted stark racial hierarchies in which some variant of white European Americans occupied the highest rung on the ladder and ethnic and racialized groups—including African Americans, Mexicans, and Asians, as well as Eastern European Jews, Italians, and Poles—were consigned to the lower rungs.[11] Today the circular and blatant racism of early-twentieth-century eugenics, which assumed that complex phenotypical and behavioral traits were the result of one gene, is easy to recognize and criticize. It is plain in the rationale behind the passage of the federal 1924 Immigration Act, which restricted the entry of "undesirable" immigrants from southern and eastern Europe and Asia, and in racial purity laws that prohibited interracial marriage in many states until the U.S. Supreme Court ruled such statutes unconstitutional in *Loving v. Virginia* in 1967.[12] Less obvious, however, are the insidious ways in which medical genetics and genetic counseling buttressed racial segregation and fostered racial differentiation, even as well-known geneticists condemned biological racism as a relic of the past.

To start, there was neither a complete nor smooth transition from the eugenic racism of the 1920s to the liberal genetics of the 1950s. Even as Mendelian eugenics, which divided races into discrete groups possessing singular inherited "unit characters," dissipated, "race" always hovered close to the surface. This persistence of race across the twentieth century is often lost under the weight of arguments that "race as culture" supplanted "race as biology" in the years following World War II.[13] Yet this is too narrow of a lens through which to account for both the recalcitrance of "race" as a biological category and the formulation of new classificatory schemes intermeshed with concepts of population, biotypes, and human constitution. From the perspective of science and technology studies, Jenny Reardon offers a comprehensive challenge to the biology-to-culture shift, criticizing the transition from "races" or "types" to "populations," from "typological" to "population" thinking.[14] Dissecting the divergence of the two earliest United Nations Educational, Scientific, and Cultural Organization (UNESCO) Statements on Race (1950, 1951) and the theories espoused by Montagu and Dobzhanksy, Reardon concludes that "the distinction between the old typological approach and the new population approach does not provide a stable foundation for assessing the history of race and science."[15] She instead shows how "race" was reconfigured through populationist ideas such as clines, ethnic groups, and geographical diffusion. Both UNESCO statements agreed with the general assertion that "race" was a category that "designates a group or population characterized by some concentrations, relative as to frequency and distribution, of hereditary particles (genes) or physical characters, which appear, fluctuate, and often disappear in the course of time by reason of geographic and/or cultural isolation."[16] In other words, "race" continued to have scientific value once detached from biological racism and appended to newer theories of populations and gene frequencies.

A close reading of *Heredity, Race, and Society*, written by Dobzhansky and the zoologist L. C. Dunn, who both helped to craft the second UNESCO statement on race, underscores Reardon's astute analysis.[17] In this tract, Dunn and Dobzhansky celebrated the biological variation of humankind, lambasted doctrines of racial inferiority and superiority, stressed the importance of environment in the development of individuals and the species, rejected bans on intermarriage, and disabused their readers of the myth of "pure races," writing that "mankind has always been, and still is, a mongrel lot."[18] In place of unequivocally jettisoning "race," however, Dunn and Dobzhansky contended that "race" could no longer be linked to phenotype or to distinctions derived through anti-quated sciences of measurement such as phrenology and craniometry. They formulated a definition of "races" as "populations which differ in the frequencies of some gene or genes," emphasizing that "races exist regardless of whether we can easily define them or not."[19]

As philosopher of science Lisa Gannett argues, "population think-ing" allowed the perpetuation of racial stereotypes largely through the prism of statistical probability, or the idea that a person from Popula-tion Z is exceedingly likely to possess the traits associated or correlated with Population Z.[20] Gannet explains, "Populations did not replace races. Races were reconceptualized as populations and a 'populational' con-cept of race was substituted for a 'typological' one."[21] The populational concept resonated very well with the idea of "racial isolates," a conflation that underpinned the mission of the Human Genome Diversity Project, an initiative attached to the Human Genome Project in the 1990s. The diversity project identified certain indigenous and ethnic communities across the globe as "population isolates" and sought to study them from the perspective of human evolution. However, these communities, work-ing with the activist Third World Network, angrily rejected this scien-tific endeavor, calling it a "Vampire Project" and a biocolonial attempt to extract biological tissue samples from their bodies and communities.[22] More generally, it can be argued that the uneasy and partial submer-gence of "race" under "population" has facilitated rather than hindered racial thinking in terms of genetic difference, whether in regard to dis-ease demographics, criminal predisposition, or innate mental capacity.[23]

Most of the early genetic counselors, such as Reed and Lee Raymond Dice of the University of Michigan, relied on the population concept, with its intractable racial underpinnings, to understand human dif-ferences and explain variation in the gene pool, and their approaches reflected the incongruous intertwining of race and population. Further-more, midcentury practitioners of genetic counseling adopted the com-panion frameworks of population control and family planning. From the 1940s to the 1960s, medical geneticists such as Reed, Dice, and Nash Herndon at Wake Forest University lectured regularly on the dangers of the population explosion with an eye toward regulating fertility patterns in the "third world." For example, in 1951 Reed discussed possibilities for "brakes on population growth" as part of the lecture series "Population Pressure versus Food Resources" held at the University of Minnesota.[24] In 1960 he spoke before the local Optimist club about the need to educate "the vast masses in underdeveloped countries to have fewer children."[25] In a 1965 article on the distribution of intelligence across different sectors of society he penned for the British *Eugenics Review*, Reed exhorted that "the need for eugenic concern is greater to-day than ever before because of the population 'explosion.'"[26] Both Reed and Dice carried on lengthy correspondence with the Population Reference Bureau and participated in population control conferences held in the United States and Europe. The studies conducted by these men and their colleagues were framed squarely by conceits about population control and the need to map gene frequencies among various delimited social, racial, and ethnic groups. Try as they might to promote the values of racial liberalism in all aspects of their lives and their work, for medical geneticists such as Reed and Dice, the persistence of racism in American society and the racialized construct of population made attaining such a goal elusive if not impossible.

Anonymous Racism?

Starting in 1910, the Eugenics Record Office, located in Cold Spring Harbor, Long Island, served as the country's central clearinghouse and training center for eugenics.[27] In 1939 it was forced to disband, in large part because the prejudiced views of its superintendent, Harry H. Laughlin, were alienating many of his colleagues and the ERO's main benefactor, the Carnegie Foundation. However, when the ERO closed its doors, new life was breathed into its mission by the Pioneer Fund. Established in 1937 with Laughlin as founding president, the Pioneer Fund was underwritten by Wickliffe Draper, a reclusive millionaire who had inherited a textile fortune.[28] Draper admired Adolph Hitler and Nazi policies and was a firm believer in the "Back to Africa" movement, which sought to relocate Americans of African ancestry across the Atlantic in a reverse diaspora of forced migration. Draper devoted his enormous resources to such causes and associated himself with scientists who appeared to be sympathetic to theories that blacks were biologically inferior.[29] Following the trail of eugenics forward into the 1940s and 1950s, Draper decided to target newly established medical genetics clinics in the hopes that they would be sympathetic to proving his highly charged thesis.

When the country's first three heredity clinics started in 1941, they straddled the basic sciences and medicine. While the in-between status of the fledgling field of medical genetics clinics meant these clinics could pursue creative interdisciplinary projects, it also meant they were marginalized institutionally and intellectually. How could the backwater of medical genetics launch substantive research into human inheritance, genetic diseases, and reproductive patterns? Then as now, funding was at a premium. Chasing the money reveals that while leading geneticists were quick to denounce the biological racism associated with eugenics, they were far more forgiving when it came to their wallets. Fully cognizant of Draper's views, two of the trio of heredity clinics accepted financial support from him in the 1950s.

Draper contacted the heredity clinics about his proposition in the 1940s. The initial reaction of the human geneticists was consternation about his ideas and intentions. James V. Neel, who had joined the University of Michigan's Heredity Clinic after serving in the U.S. Army and directing the Atomic Bomb Casualty Commission, exchanged many letters regarding Draper with both Reed in Minnesota and Herndon in North Carolina. Neel's position was that Draper threatened the scientific and ethical development of medical genetics. He told Herndon, "My first inclination is to have nothing to do with this gentleman," adding, "I have a very healthy respect for what his millions could do to the study of human heredity in this country if misdirected. The question is, to what extent can one sacrifice present principle for a possible later good." Neel expressed that he had "too much at stake in the field of medical genetics to get tinged with a 'racist' label."[30] Throughout his career, Neel stood out among his colleagues for regularly rebuffing the overtures of the American Eugenics Society.[31] For example, in 1953, Neel diplomatically declined one of many requests to join from the society's president, Frederick Osborn: "I cannot help but feel that the term 'eugenics' by common usage has connotations with which I am not in agreement. Accordingly, I think that for the present I shall continue my own efforts to advance our knowledge of heredity in man outside the framework of the American Eugenics Society."[32]

In the end, however, Neel and his colleague Dice blessed a $100,000 gift from Draper. After a series of forthright conversations with Draper,

Neel and Herndon became convinced that the philanthropist would permit them to study crucial human differences outside of a racist black-versus-white dichotomy. They decided that Draper's munificence could be applied to analyzing mating patterns in endogamous white communities. This compromise between Draper and geneticists showed how hypotheses about human differentiation could be redirected from inter-racial to intra-racial differences. The family studies conducted by eugenicists in the late nineteenth century and early twentieth century had focused primarily on European Americans, usually showing the supposedly degenerate lineages of poor, rural, white, and Southern families.[33] By the 1920s, eugenicists had racialized degeneracy, as immigrants from eastern and southern Europe, Asia, and Latin America became the biologically inferior menace. However, as the endeavors underwritten by Draper suggest, the logic of human differentiation was malleable, now attached to population concepts of "racial isolates" and "clines," whose mating and reproductive dynamics demanded scientific scrutiny.

At the University of Michigan, this translated into a study of "assortative mating in a city community." Conveniently choosing Ann Arbor as the living laboratory to gauge the "trend of heredity in a modern human population," Dice intended to test the supposition that people with similar attributes tended to marry each other, in this case among the educated white middle class.[34] This project would help Dice enhance his profile as a human geneticist—not just a laboratory or field biologist with expertise in small mammals. In spring 1950, Dice accepted Draper's advances, soon receiving $100,000 to be allocated over a five-year period.[35] Seeking discretion, Draper requested, and Dice consented, that his gift be treated as anonymous.[36] By fall 1950, Dice had hired a research associate, James N. Spuhler, to conduct the fieldwork for this longitudinal study, which resulted in an article published almost two decades later in the *Eugenics Quarterly*.[37] For his part, Neel must have been relieved that his name would not be associated with Draper or the Pioneer Fund. Yet he was pleased with the funding, writing to Reed in March 1950 that the Heredity Clinic "would benefit in the not too distant future from Col. Draper's largesse. We are setting up a rather good sized project on assortative mating which we sincerely hope can be kept entirely out of the realm of racist problems."[38]

Whereas Neel was cautious about Draper and resolute that no explicit strings be attached to any of Michigan's research projects, Wake Forest University's Herndon was a more willing player. In 1953 he coordinated a $100,000 gift to endow a professorship in medical genetics, which he eventually held, at the Bowman Gray School of Medicine.[39] Herndon knew well that Draper was "preoccupied with ideas of 'superiority' of the Anglo-Saxon 'race'" and told fellow human geneticist Laurence Snyder that "it would probably be justifiable to classify him as a 'racist.'" Nevertheless, Herndon coveted support for his department and research projects and thus presented Draper with "a series of half a dozen or so interrelated projects involving some fairly extensive gene frequency studies in the various population groups available in this state." Draper expressed immediate interest in studies of the "population of the Blue Ridge and Smoky Mountains" and was willing to underwrite a "study on the mating structure of this population, including factors affecting assortative mating, size of isolates, etc." In return for the money, Herndon assented to Draper's two conditions, namely, that his "department not advocate miscegenation," and "that sterilization be accepted as a therapeutic weapon where medically indicated."[40] Neither of these gave Herndon any pause.

It is not shocking that Wake Forest's medical genetics department was the most amenable to Draper's agenda. Located in Winston-Salem, North Carolina, the program was founded by William Allan, a general physician interested in hereditary diseases and eugenic prevention who had moved his family records office from Charlotte thanks to a five-year start-up grant from the Carnegie Foundation.[41] When Allan died unexpectedly in April 1943, Nash Herndon, with whom he shared a family pedigree as a second cousin, became the long-standing director and sustained Allan's vision.[42] As the blueprint for the program stated, the "geographical location" of Wake Forest was "highly favorable" because the population was "stable and relatively homogeneous, and predominately rural." In other words, what they in less formal writing referred to as the inbreeding poor whites of the Smoky Mountains held "great genetic advantage" as a kind of unique American racial isolate of "post-Revolutionary pioneers" that did not include any "negroes, tenant farmers or shifting mill population."[43] Allan and Herndon viewed their environs as a "laboratory" that should be exploited "to obtain the information concerning gene frequency, mating isolates, etc., which would permit the intelligent development of large scale programs for prevention of hereditary diseases." The region was the ideal site for the "application of eugenic programs" because there was "practically no opposition in this area to such programs."[44] After one year Allan reported that much progress was being made: "In the Smokies the sheep can be separated from the goats, and a good start has been made toward solving this most difficult genetic problem."[45]

During the early 1940s, Allan and Herndon initiated the Watagua County Survey, which sought to study patterns of disease inheritance in a population characterized by poverty, inbreeding, and ignorance. They also embarked on surveys of crippling diseases and hereditary blindness. Allan was particularly interested in a eugenic program that would detect carriers of recessive genes—such as albinism—so that they could be advised against reproduction.[46] Wake Forest's medical genetics department also furthered North Carolina's sterilization program, which was one of a few in the country that maintained county eugenics boards and that recorded an increase in operations from the 1940s to the 1960s.[47] Specifically, Allan and Herndon started the Forsyth County Eugenics Program, which collaborated with the local county health officer in a "gradual, but systematic effort to eliminate certain genetically unfit strains from the local population."[48]

In the end, Reed was the sole decliner of Draper's conditional generosity. Given Reed's strong racial liberalism, it makes sense that he would have been the most vocal in refusing Draper's gifts. Expressing his opinion about Draper, Reed wrote to a friend, "Colonel Draper has very definite ideas as to what the subject of human genetics encompasses," foremost among them the "improvement of the American people by shipping the Negro inhabitants back to Africa."[49] After receiving an inquiry about Draper's agenda, Reed stressed that he never wanted to be quoted in reference to Draper, who "approaches the laboratories in a bombastic fashion and I am sure that it is very difficult to get any money out of him which one could use in any way that would be sensible to them. . . . He did not really know any genetics himself and was a racist of the usual type."[50] Reed also explained that he never romanticized the many trophies from African safaris that hung in Draper's lavish New York apartment and thought of him as an eccentric bachelor on a misguided crusade. Unlike fellow geneticists Dice, Neel, and Herndon, Reed maintained strict distance from Draper. Nonetheless, Reed did indulge the bounteousness of Charles M. Goethe, the founder of the Eugenics Society of Northern California and an unrepentant chauvinist who celebrated Nordics and Anglos and especially disparaged Mexicans.[51] Reed benefited from the regular infusions of small sums that Goethe sent to a handful of human geneticists with eugenic leanings for purposes that ranged from conference travel to book purchases to buying Thanksgiving turkeys for up-and-coming junior faculty members. Reed cultivated such a good relationship with this grandiloquent Sacramento magnate that in his final years Goethe decided to bequeath almost a half million dollars to the Dight Institute, which the University of Minnesota began to receive in installments after his death in 1966.[52]

Adoption and Racial Matching

Genetic counselors worked to uphold the racial order in mid-twentieth-century America by serving as experts for child welfare agencies attempting to match relinquished newborns to adoptive families.[53] During the heyday of domestic adoption in the United States from the 1940s to the 1960s, heredity clinics were inundated with requests from welfare and children's agencies to evaluate infants for skin color, racial characteristics, medical problems, and the likelihood of a genetic disorder.[54] In helping to forge American families through adoption, genetic counselors such as Reed upheld the doctrine of familial homogeneity and in turn helped to draw firm racial boundaries. At the time, the convention was that parents and child should look as similar as possible and that phenotypic similarity created an "almost mystical notion of identification: attachment between those who look and therefore are alike just happens."[55] With few exceptions, most adoption and welfare agencies followed the one drop rule: a child with any appreciable "Negroid ancestry" could only be placed in a nonwhite family.

In the 1940s and 1950s the single most common reason for genetic counseling was to evaluate a newborn for adoptive placement. In his 1955 *Counseling in Medical Genetics*, Reed listed the top twenty causes for consultation at the Dight Institute. First was skin color, followed by epilepsy, consanguinity (usually cousin marriage), mental deficiency and mongolism, schizophrenia, and 14 more conditions.[56] When the Dight Institute reopened its doors under Reed in 1947, adoption cases predominated. One involved a local child welfare bureau interested in the adoptability of a "near white" baby and the "usual question as to whether his [the baby boy's] children could show prominent Negro characteristics."[57] The state division of social welfare presented the institute's second adoption case, a 4-year-old girl with some Negro ancestry who was going to be placed for adoption, inquiring about the likely racial appearance of her offspring, in other words, the prospective adoptive parents' future grandchildren.[58]

The most common reason for an adoption consult was not to determine the existence of a genetic disorder but to detect any possible trace of "Negroid" ancestry. In a survey of adoption assessments performed at the Dight Institute from September 1947 to December 1957, Reed reported that 73 of a total of 165, or 44 percent, involved assessment for a "racial cross known or suspected," with 54 or 74 percent of these for "Negro," and the remainder for American Indian, Italian or Mediterranean, Japanese or Oriental, Mexican, and Filipino.[59] Further down on the list were consultations about neurological or mental defects in a relative, children born to a consanguineous or incestuous mating, congenital malformation in a relative, and miscellaneous queries.

Typically, adoption and welfare agencies wanted to know about a newborn's eventual ability to convincingly appear white in American society. Reed employed several diagnostic criteria to determine if a newborn would be able to "pass for white" and "thus enjoy the better socioeconomic conditions of the white community."[60] He examined the sacral spot at the base of the spine, which is never "seen in blond children with blue eyes," finger smudges on the backs of fingers between the joints, skin color (ideally when the newborn was at least 6 months old), nose width, lip thickness, and hair shape and texture, and he looked for an epicanthic eye fold common among so-called mongoloids.[61] Despite his statements dismissing the importance of skin color, Reed regularly referred to the University of California at Berkeley geneticist Curt Stern's hypothesis that there were four or five pairs of skin color genes.[62]

In March 1950, a caseworker from the local Children's Home society brought in a baby to ascertain whether it had "negro blood." Reed classified the newborn as having a thick nose, coal black curly hair, colored scrotum, two large Mongolian spots, and pigments on the backs of fingers. Using these criteria, he thus "advised that it be considered colored."[63] In another case, the prospective adoptive parents were concerned that a baby boy they called Richard might manifest "Negroid characteristics." Although his biological mother and her ancestors ostensibly were all white, his biological father was the "son of a Negro woman and a white man." The potential parents expressed a fear common in the 1940s and 1950s, namely, that future offspring could revert to "Negroid," sometimes referred to as an atavistic "throwback."[64] Reed sought to dissuade his clients of this erroneous notion, and in Richard's case he reassured the adoptive parents that "there will be no reversion to character-istics any more negroid than those which Richard may possess."[65]

Reed believed that racial matching was a good concept since it pre-vented "embarrassment from arising which would result from an adopted child with traits which could not possibly have come from the adoptive parents."[66] In other words, racial matching meant adhering to normative patterns of family and parent-child appearance in midcentury American society. Ultimately, Reed cared most about the well-being of his white middle-class clients and not the feelings of mixed-race parents, who in many states were unable to marry because of "social prejudice and pres-sure against a marriage which would have provided legitimacy." Instead of worrying about the fate of unmarried mixed-race couples whose chil-dren he deemed to be the "best adoption risks," Reed sought out a larger pool of white foster parents that were "free of racial prejudices and also match[ed] the children to some extent in appearance."[67] According to Reed, if these ideal conditions were met, the adoption was "expected to be highly successful," even though "parents willing to adopt such a child must always be ready for the appearance of a "dash of 'colored blood.'"[68]

Even as Reed reinforced racial homogeneity among American fami-lies with his recommendations for placement, he viewed mixed-race children as the best prospects for adoption. Based on his work at the Dight Institute, Reed had identified three categories of potentially avail-able children. The first were legitimate children from broken homes, who faced difficulties and carried "an appreciable risk" of having "genes for mental and physical deficiencies."[69] According to Reed, these were the poorest prospects for adoption. Slightly better but still not ideal were illegitimate children who might be of good genetic stock but perhaps only of poor social conditions and hence were very much in demand and in short supply. The final group, children of mixed racial ancestry, Reed believed to be the "most vigorous and healthy stock generally available for adoption."[70] Yet the demand for such children was low. From Reed's perspective, couples unable to conceive had various options to create a family in a context characterized by a dearth "of newborn blue-ribbon babies." In need of adoption were older children and those "with some percentage of African or Asian ancestry." Given these suboptimal choices, Reed stated that the "products of racial crosses are the best adoption risks that there are."[71] The seeming contradiction of Reed's simultane-ous sanguine evaluation of mixed-race children and his insistence of racial matching illustrates the extent to which he could escape neither the scientific theories of racial differentiation he sometimes so vocally condemned nor the expectations of phenotypic familial similarity that predominated at the time.

Adoption consultations were also weekly fare at the University of Michigan's Heredity Clinic, which filed 111 cases from 1942 to 1971 under the label of "Adoption."[72] Of these 111 Adoption files, 45 cases (41%) were categorized as "racial ancestry" or "racial characteristics" inquiries. The remaining 66 cases (59%) covered a wide array of conditions that changed over time in tandem with developments in the identification of genetic disorders and genetic tests. The most common among the medical conditions included neurofibromatosis, Huntington's disease, cerebral palsy, epilepsy, muscular dystrophy, retinis pigmentosa, incest (father/daughter and brother/sister), mental retardation, short stature, and schizophrenia. In addition to the Adoption files, there was a separate set of 129 files labeled "Racial Characteristics" that dated from 1942 to 1977. Of these, 39 were cross-referenced with the Adoption files, but at least 54 were additional inquiries related to adoption.[73] In total then, the Heredity Clinic handled approximately 100 racial matching/adoption cases from the 1940s to the 1970s, the majority from the mid-1950s to the mid-1960s. These cases came from the Michigan Children's Institute, located in Detroit; Catholic, Protestant, and Jewish agencies throughout the state of Michigan; and probate courts and children's hospitals. All but two of the inquiries came from a city or county in Michigan, covering the state from north (Traverse City) to south (Kalamazoo) and from east (Detroit and Saginaw) to west (Grand Rapids). They illustrate the integral role that medical geneticists played in adoption in Michigan during the mid-twentieth century.[74]

The University of Michigan's Heredity Clinic frequently called upon a colleague from the anthropology department to participate in the assessment. As Neel explained, in cases where the Heredity Clinic was asked to ascertain racial ancestry, he assembled a panel consisting of "two anthropologists and myself." The panel did not seek to answer "the question of whether there is any Negro ancestry, in a case such as you describe, but only whether the child's appearance is such that he would most appropriately be placed in a Negro home or in a Caucasian home. As you are well aware, there has been a certain amount of Negro admixture with Caucasians, particularly in southern Europe."[75]

Because skin color changed during the first months of infancy, Neel, who oversaw many of these cases, frequently told agencies that his team

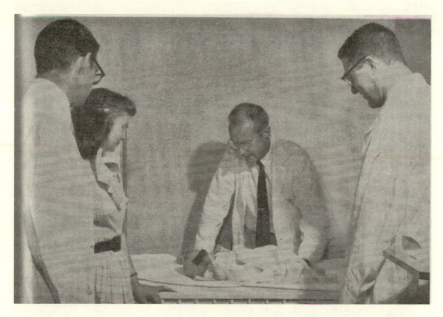

A team of two physical anthropologists, one social worker, and one medical geneticist assessing the racial classification of a newborn at the University of Michigan's Heredity Clinic, circa 1957. "The child on the examining table is illegitimate, and it is known that his parents are of different races. Is this child to be raised and brought up with a Negro family? Or is he to be raised as a white child?" *Source*: Charles F. Wilkinson Jr., "Heredity Counseling: Genetics Clinics and Their Preventive Medical Implications," *Eugenics Quarterly* 4, no. 4 (1957): 205. Reproduced by permission.

could not adequately assess a baby's racial ancestry until it was at least 6 months old. As he explained to a caseworker at Detroit Catholic Social Services in 1959 regarding a 3-month-old girl,

> We prefer not to be called upon to express any opinion before a child is six months of age. This is because during the neonatal period the physical characteristics, including the pigmentation, of a child undergo rather rapid change, so that a much more valid opinion can be rendered by the time a child has reached six months of age than at, say, two months of age. While I appreciate the strong desire on your part for an early decision, to facilitate plans for the child's placement, I believe that in this particular case it might be wise to delay placement somewhat beyond the usual early age.[76]

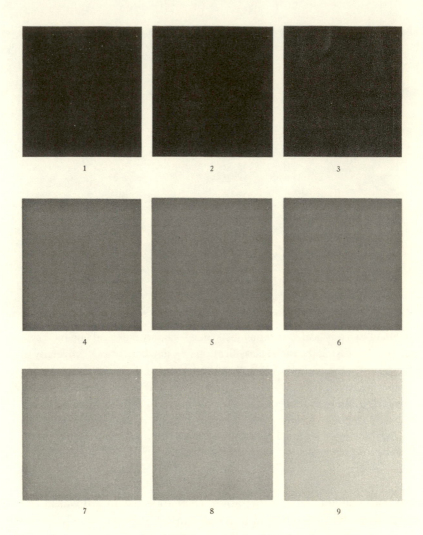

Each rectangle, numbered 1–8, represents the skin color of an individual colored person. No. 9 is the skin color of a white person (the author).

From the 1940s to the 1960s the heredity counseling team at the University of Michigan used the Gates color scale (reproduced here in black and white), developed by R. Ruggles Gates, to match newborns with racially compatible families. *Source*: Front insert in R. Ruggles Gates, *Pedigrees of Negro Families* (Philadelphia: Blakiston, 1949).

He asked the case worker to contact him again once the girl had reached 5 months of age.

How did Neel and his colleagues racially classify a newborn? The Heredity Clinic's method was more streamlined and employed more criteria than Reed's at the University of Minnesota. Neel evaluated infants according to physiognomic traits: hair (color, straight or curly, fine or thick); nose (infantile, broad); ears (large, free or attached lobe); skull (mesocephalic, minor frontal bossing); eyes (color); lips (thickness, including of lip seam). Most importantly, Neel and his colleagues used a scale developed by Reginald Ruggles Gates to rate skin pigmentation on a scale of 1 to 9, with 1 being the darkest and 9 the lightest.

Gates was a Canadian biologist born in 1882 who studied plant genetics at McGill University and the University of Chicago, where he received his PhD in 1908. Gates focused on the emerging field of cytology and specifically the study of chromosomes and mutation. By the 1920s, he was applying his genetic knowledge to humans, and in 1923 he published the book *Heredity and Genetics*. He then began to extrapolate from plant to human genetics in regard to racial differences, which he asserted "could not be accounted for in terms of single gene pairs."[77] Gates proposed that a small number, two to five gene pairs, were responsible for racial characteristics, including skin color, ear size, and whether the ear lobe was attached or detached. Underlying his views were strong racial prejudices, particularly against miscegenation or racial mixing between whites and blacks. In 1947, Gates held a fellowship at the historically black Howard University in Washington, D.C., where 18 academics signed a petition asking for his dismissal.[78] Gates was noticeably at odds with the revisionist scientific proclamations about race that were being made by UNESCO-affiliated scientists in the late 1940s and early 1950s, during which time he was a contributor to the racist journal *Mankind Quarterly*. Despite Gates's problematic career, Neel and his colleagues relied extensively on the Gates Scale.

In 1957, a revealing case labeled "Racial Characteristics" came to the Heredity Clinic. Less than two years earlier, when she was 27 years old, Gertrude was walking home to her own apartment at 12:45 a.m. from a girlfriend's house after watching television. A working-class young woman, Gertrude was born and raised in Detroit, by a father who worked for a brewery and a mother who was a housewife. She completed schooling up to the ninth grade and then began working as a waitress. By her early 20s, Gertrude had a job as a sorter for Prince & Company, which printed automobile news magazines. Near an alley less than one block from her home, two men grabbed Gertrude; one muffled her mouth while the other man raped her. Because it was so dark, she could not see her assailants. She was too scared to file a police report and soon found out that she was pregnant. Eventually she phoned Kiefer Hospital in Detroit and made arrangements for prenatal care and the eventual adoption of her child. She only told one of her brothers what had happened and expressed great shame about the event.[79] The county juvenile agent wrote the report about Gertrude's rape, noting that in February 1956 she gave birth to a healthy baby boy who was ready for adoption placement. However, the boarding home mother for the Children's Institute in Kalamazoo was convinced that the "facial features of the Negro race" were "evident although the child has light brown hair, fair skin and blue eyes." The county juvenile agent himself did not see evidence of Negroid features but was concerned because the police involved in this case had reported that "a lot of men of that race" were "in that neighborhood."[80]

Perhaps Gertrude's Italian ancestry played a role in the assessment of her as less than lily white, but in any case, these suspicions of the birth father's racial background prompted the county juvenile agency to contact the University of Michigan's Heredity Clinic to determine if this boy, now over 1 year old, had any "Negroid ancestry." According to the clinic's expert assessment, he did not, presenting with fine straight hair, blue eyes, and "very light" skin pigmentation "less than #9 on the Gates scale, without mongoloid spot." Neel explained that the panel judged that there was "nothing about the child's appearance to suggest this possibility" and recommended that he be placed in a "Caucasian home."[81]

In another case in 1960, Catholic Social Services of Wayne County referred a 4-month-old boy to the Heredity Clinic for an assessment of adoptability. The boy was a legal ward of the state, and his mother was a patient at the Wayne County Training School, classified as feebleminded with a reported IQ of 49.[82] Although the boy's mother was "white, French, olive complected" and he was "so white that he would readily pass for white had not the question been raised by his supposed background," a pediatrician involved in the case was concerned about the boy's potential adoption placement because of suspicions he was conceived "during a period when the mother ran away from a home placement and lived with a negro man for several days."[83] Given the infant's situation, the agency wanted the Heredity Clinic's "advice as to whether you think it would be advisable for him to be placed with a white or colored family."[84] Led by geneticist Margery Shaw, the Heredity Clinic's expert panel of three evaluated the boy the following month and noted that he had "sallow, yellow, but fair skin color (#9 on Gates' scale), dark brown and slightly wavy hair with fine texture, brown irides, medium lips with medium lip seam, free ear lobes, broad, flat, infantile nose, and no excessive nipple or sacral pigmentation." While Shaw's team deemed the infant "well within the range of 'normal' for the Caucasian race," they cautioned Catholic Social Services that the Heredity Clinic could not make a more accurate determination until the child was close to 1 year of age. They also noted that a family history of feeblemindedness could present a "greater deterrent" to adoption than any questions about his skin color.[85] Close to one year later, this child returned to the Heredity Clinic for follow-up. Concerns about his mental aptitude were soon put to rest as he scored 115 on an IQ test and was ranked as "bright normal" based on the Cattell Infant Intelligence Scale and the Gesell Development Schedules. Again scoring #9 on the Gates Scale, the Heredity Clinic panel recommended "permanent adoptive placement with a Caucasian couple," noting that "there seems to be no genetic contra-indications to adoptability."[86]

Into the 1970s caseworkers from agencies across the state requested adoption evaluations from the University of Michigan's Heredity Clinic. For example, in 1970 the Child and Family Services of Saginaw asked for the assessment of a 3-year-old girl whose adoptive parents wanted "predictions as regards to the racial characteristics of her children should she have any." Neel concluded that this toddler did not appear to have "any appreciable Negro ancestry" given her fine, straight, and straw-colored hair; blue eyes; infantile broad nose; and very light skin pigmentation. Seemingly the adoptive parents were concerned that the child's biological father had measures of "Negro ancestry," a possibility that Neel did not find, reiterating that "we discounted once again the old folklore about the 'throw-back.'"[87]

By the 1960s, like Reed, Neel and his colleagues began to express the hope that open-minded white parents, especially those residing in more racially and ethnic diverse communities, would adopt light-skinned biracial children. In June 1967, for example, a Heredity Clinic panel including a social worker, one physician, and one medical geneticist saw a young boy who apparently had been adopted by two white parents. The child's traits were noted as brown hair with loose curls, dark brown eyes, moderately full lips, a flat-bridged nose with flare, and a pigment of #6–7 on the Gates Scale. They classified him as a fairly light-skinned

African American boy yet stated that he possessed "sufficiently Negroid physical traits that he is unlikely to pass as Caucasian. We do not make any attempt to estimate the percent of racial ancestry. We feel that [his] future welfare and adjustment in society will depend more directly on the attitude of his adoptive parents than on his non-Caucasian traits."[88]

This liberal position was gaining currency in American society. Pearl Buck, for example, mother to both a girl with developmental disabilities and an adopted girl, strongly endorsed interracial adoption and called on accepting white parents to give a home to relinquished "brown babies."[89] As for Reed, after retirement he worked as a volunteer, helping to resettle the many Hmong families that came to Minnesota fleeing violence in Southeast Asia at the end of the Vietnam War. Reed learned the Hmong language, and his wife cooked for these exiled families. In the 1990s, one Hmong student attending community college, Yeng Yang, lived in the Reeds' basement, surrounded by the African violet collection that Reed crossbred with much affection according to the principles of plant genetics. Yang told a local reporter writing a story about the couple's community involvement that Reed had changed his life and he considered him a father.[90]

Racial Legacies in a Multiracial Society

In the early 2000s, Wake Forest apologized for its eugenics program and for accepting Draper's money.[91] This story briefly made sensational news in North Carolina. Yet it is important not to let the distasteful extremes of the story of Draper and the Pioneer Fund impede us from seeing the ways in which racial differentiation operated, much more mundanely, in the wider realm of genetic counseling. Most of the early practitioners of genetic counseling, such as Reed and Dice, strongly supported population control, family planning, and the objectives of organizations such as the Population Council. These projects, while sometimes engaged in expanding access to birth control in underserved communities, were racialized endeavors that tended to demonize overly fecund families in third world countries and poor communities in the United States that exceeded the ideal of 2.1 children per household. This logic of population control—that certain kinds of people needed to reduce breeding and make use of contraception—was pervasive in the Cold War decades of the 1950s and 1960s and wielded as an antidote to Communism.[92] The medical abuses linked to racism and population control policy exploded onto the scene in the 1970s when lawsuits and allegations of the forced sterilization of women of color and poor white women appeared in newspapers, congressional hearings, and court rooms. These developments were crucial blows that facilitated the final dismantling of the eugenics movement in America, which had been partially reconstituted under the umbrella of population planning after World War II.[93]

By the 1970s patterns of family formation were changing, with the decriminalization of abortion and growing openness to biracial families at least in some parts of the country. These shifts, along with rising rates of international adoption, helped to bring about a decrease in domestic adoptions and concomitantly the placement consultations that for so long were core services at heredity clinics. The master's-degreed genetic counselors of the 1970s encountered race and ethnicity not by participating in the racial classification of specific individuals but by partaking in newly launched screening programs aimed at particular groups, most notably for Tay-Sachs disease in Ashkenazi Jewish communities and sickle cell anemia in African American communities.[94] Scholars have explored insightfully the extent to which the eugenic ghosts of doctrines of racial and ethnic superiority and inferiority infused early genetic screening programs even when they began as empowering grassroots community efforts.[95] These genetic public health endeavors also raised the question of the advisability and ethics of identifying carriers of recessive genes for the sole purpose of regulating reproduction, not for the diagnosis of a potentially treatable disease condition. Today, genetic counselors incorporate ancestry—as determined by various modalities of population genetics—into their work when they identify individuals and families for genetic testing based on either self-reported ethnic or racial classification or, less frequently, entire genome sequencing. Although this might seem inoffensive on the surface, genetic counselors practice in a society where associations between medical genetics and racism have not evaporated in the twenty-first century. In the minds of some Americans, especially African Americans and Mexican Americans, genetic testing and the prospect of genetic counseling conjure up fears of medical mistreatment or memories of coerced sterilizations in the 1960s and 1970s.[96]

While there appears to be considerable amnesia about genetic counseling's racialized past, some genetic counselors have sought to address these historical legacies head-on. Over the course of the more than two decades that Nancy Steinberg Warren directed the genetic counseling program at the University of Cincinnati, she became convinced that the only way to diversify her profession was to infuse discussions of race, class, and culture into "every course in every quarter in every clinical rotation." She arrived at that conclusion after studying the conundrum of genetic counseling's demographic homogeneity from many perspectives, some of which took her and her colleagues out of their comfort zone. Warren captures this dynamic poignantly when describing a two-day retreat she organized in 2004 to explore issues of diversity (or lack thereof) in genetic counseling. A small but engaged group of about 25 participants with various affiliations gathered in Dayton, Ohio, to tackle the topic. The retreat's first day began with several introductory lectures on genetic counseling. About midmorning, one of the diversity representatives raised her hand and said, "We've been talking about genetic counseling for over two hours, but I still don't know what you do, I don't know why somebody from my community—Native Americans in rural communities—would want to be a genetic counselor. You need to reframe how you are talking about genetic counseling." This participant elaborated, explaining that what her community knew about genetics, namely, eugenics and sterilization, was not very nice and could not be disregarded as mere misconceptions. She also articulated the theme of giving back to one's community, which rarely had been incorporated into genetic counseling recruitment materials. The candid and "heart-wrenching" conversations that unfolded at this meeting made Warren cognizant of the enormous challenge faced by genetic counselors that view diversification as paramount to the profession's survival in and relevance to multicultural America.[97] With this aim in mind, Warren, thanks to a grant from the National Society of Genetic Counselors, created a Genetic Counseling Cultural Competence Toolkit to provide accessible, multicultural classroom, clinical, and community outreach resources.[98]

Along these lines, other genetic counselors are engaged in research projects that seek to understand why it has been so difficult to branch out in terms of race, ethnicity, and gender. These studies suggest that avenues for change include reaching potential genetic counselors at the high school or middle school level, ensuring that basic medical genetics is part of science education in high school and college, developing programs to encourage recruitment and retention of minority genetic counselors, and working closely to build trust and alliances with community health organizations that serve minority and poor populations.[99]

CHAPTER FOUR

Disability

The Dynamics of Difference

I n December 1961, WBAL, the NBC affiliate in Baltimore, aired *The Dark Corner*.[1] A local production that took nearly two years to prepare, this documentary exposed the shrouded and difficult lives of people with mental retardation in Maryland. Released to coincide with President John F. Kennedy's announcement of the creation of an expert panel on mental retardation, *The Dark Corner* drew heavily from the medical expertise and personal experience of pediatrician Robert Cooke, head of the Harriet Lane Home for Invalid Children at the Johns Hopkins Hospital and father of two severely disabled girls with cri du chat syndrome.[2] Like President Kennedy's sister Eunice Shriver, Cooke wanted to upend the shame and stigma associated with mental retardation. Along with a vocal set of parents, physicians, and legislators, Cooke sought to bring the issue of mental retardation to the national agenda. He applied the language of human rights and the liberal Christian "beloved community" to intellectual disability in his writings, speeches, and media appearances.

In *The Dark Corner* Cooke reflected on caring for his two girls, neither of whom could talk or walk.[3] A house officer at New Haven (Yale University) Hospital when his eldest daughter, Robin, was born, Cooke and his wife ignored the tide of advice and opted against institutionalizing her or her younger sister, a decision that allowed them to gain a deeper understanding of their children as emotional and physical beings.[4] A compelling man who spoke with equal parts authority and sensitivity, Cooke denounced the eugenic approach to mental retardation, which encompassed "solutions" such as sterilization or long-term institutionalization and glossed over important distinctions in diagnoses and clinical manifestations. As Cooke explained in *The Dark Corner*, "It seems much more reasonable to improve the cultural environment for such individuals, rather than relying on sterilization as a control."[5]

Cooke was at the forefront of social and attitudinal changes to physical and mental disability in America. In the 1960s and 1970s, a series of revelations about abysmal conditions and overcrowding at institutions for people with mental retardation set off a firestorm of criticism, not only of the farce of public custodial "care" but also of the inflexible separation between the normal and abnormal, abled and disabled. Americans learned that people who were physically and mentally disabled had served as subjects in medical experiments, including the infamous case of deliberate hepatitis infection in hundreds of children at the Willowbrook State School in Staten Island, New York.[6]

The era of viewing people with intellectual disabilities as a menace that requires institutionalization and deserves social exclusion is over. Today, an increasingly prevalent perspective is that they are contributing citizens with developmental limitations who can flourish from social inclusion and by residing at home or in a homelike community setting.[7] The key social actors responsible for this changed outlook have been adults and children with disabilities and their parents, as well as pediatricians, psychiatrists, geneticists, genetic counselors, legislators, and activists from religious and civic organizations. Like Cooke, they have participated in a complex process of the labeling and relabeling of disease and difference, which is deeply entangled with powerful perceptions about identity, community, and human worth.

Disability historians have written surprisingly little about the historical aspects of medical genetics and genetic counseling, although disability scholars have taken medical genetics to task for promoting negative stereotypes of the disabled and assuming that the prevention model of disease should apply to disability.[8] Much of this criticism has focused on the message and impact of prenatal screening and testing on people with disabilities and their families. Scholars of medical genetics and, to a lesser extent, genetic counseling have examined the persistence of eugenic assumptions in relation to disability and abnormality, arriving at different conclusions. Some trace a sinewy thread of the eugenic prevention of disability or perceived disability across the entire history of medical genetics and genetic testing, while others suggest that the wide spectrum of disorders placed under the umbrella category of disability, ranging from mild to severe, blurs the lines between controversial genetic selection and the justifiable anticipatory eradication of a devastating disease. A few historians and bioethicists argue that reproductive choice, when exercised autonomously by mothers and parents, ultimately trumps any broader social or moral responsibilities or concerns related to people with disabilities and the disability community.[9]

Although it has received scant scholarly attention, the development of genetic counseling has been and remains central to the multifaceted interplay of medical genetics, bioethics, and disability in America. From the 1940s to the 1960s genetic health professionals helped to set the parameters for the identities and labels that eventually informed and galvanized the intellectual disability rights movement. Paradoxically, genetic counselors laid critical groundwork that helped to build networks for subsequent, and sometimes countervailing, efforts aimed at normalization and mainstreaming.[10] By the 1970s and 1980s, as attitudes toward intellectual disability began to change in dramatic ways, the central figure in genetic counseling was no longer the self-appointed male medical geneticist sitting behind a desk at a heredity clinic, but a female professional with a specialized master's degree and board certification in genetic counseling. These new health professionals entered reproductive and genetic health as it was being redefined by disability activism and advocacy, the acceptance and application of bioethical principles, the decriminalization of abortion, and ever-expanding reproductive and genetic technologies and tests. With further advances in genomic medicine over the past several decades, the relationship between genetic counseling and disability has become more, not less, complex and challenging.[11]

Disability at the Dight Institute

When Sheldon Reed coined the term "genetic counseling" in 1947, he was one of a handful of medical geneticists making the switch from studying inheritance patterns in mice and flies to the embryonic field of human heredity.[12] According to V. Elving Anderson, his close colleague and successor at the University of Minnesota, Reed made a "sharp turn" from the *Drosophila* "population bottles" he was designing at Harvard University when he accepted an offer to jump-start the languishing Dight Institute.[13] Reed described his move from New England to the Midwest as both motivated by a desire for career advancement (institute director with tenure versus assistant professor) and "at least partially rooted in altruism."[14] Reflecting on his decision in 1970 in a letter to historian of medicine Kenneth Ludmerer, then a fellow at the Johns Hopkins University Institute of the History of Medicine conducting research for one of the first histories of American eugenics, Reed explained, "When I was teaching beginning genetics at Harvard University it gradually became clear to me that human genetics was more important than the mouse and *Drosophila* genetics which I had been working on," adding that he "was interested in the genetics of behavioral traits, such as mental retardation and the psychoses."[15] Like Lee Raymond Dice of the University of Michigan's Heredity Clinic, Reed became the director of a very new entity on the academic scene—a medically oriented clinic based in the biological sciences but not yet part of the medical campus.

The institute was the result of a bequest from Charles Fremont Dight, the founder of the Minnesota Eugenics Society, a physician trained at the University of Michigan, and an eccentric reformer who, among other things, was an alderman elected on the Socialist ticket, introduced milk pasteurization to Minneapolis, and lived for many years in a treetop house adorned with mottoes over the doors like "Progress is Man's Solution."[16] Dight became attracted to eugenics in the 1910s as it attained greater visibility in the United States. By the 1920s he was the dynamo of eugenics in Minnesota, where he advocated vigorously for the state's sterilization law, passed in 1925.[17] During these campaigns, the childless bachelor Dight realized that his estate could be left to the future betterment of humankind if he underwrote a "eugenics research and education" institute. After negotiations with the president of the University of Minnesota, his plans were finalized in 1927. The Dight Institute was established posthumously in 1938 and began operations on July 1, 1941, located in the basement of the Zoology Building.[18]

The institute's first director was Clarence P. Oliver, a zoologist. According to Oliver, the institute sought to "carry out genetic and eugenics programs by collecting information about human traits, analyzing and interpreting the data collected, and making the information available to interested persons." Oliver was not shy about stating that the institute's principal purpose was to implement a "eugenics program intended to decrease the number of defective children."[19] One of the primary ways in which Oliver implemented his eugenic vision was through promoting a brand of genetic counseling that would be anathema today. When seeing a couple with a family history of developmental or physical disabilities that might be hereditary or that already had produced a child with mental retardation, Oliver thought that the genetic counselor should pressure the parents to cease procreation, because it was "unfair to a child to start him out in life with a handicap to compete in our society."[20]

Nevertheless, with the specter of World War II and Germany's Final Solution in the background, Oliver was savvy enough to realize that enterprises that showcased eugenics could tarnish the ostensibly scientific mission of the institute. In a vision statement, Oliver explained that the institute "should limit its eugenics program to consultation with persons who have immediate genetic and eugenic problems" and that "an active program by the Institute at this time to bring about legislation for the sterilization of groups or members of families, or an intensive program of propaganda of that sort, would cause the Institute to lose the public support and would make it very difficult for us to follow a research program in human genetics and eugenics"[21] (underlined in original). To circumnavigate the problems associated with using the word "eugenics," Oliver determined that the best way to pursue his desired research agenda in Minnesota was to create a parallel organization that did not use the label "eugenic" and was not affiliated directly with the university or any other public institutions.[22] Accordingly, in 1945 he facilitated the founding of the Minnesota Human Genetics League, Inc., A Society for the Promotion of Population Research and the Improvement of Human Inheritance.[23] Over the coming decades, this league provided institutional and financial backing for human genetics studies and sponsored talks and conferences on population and fertility planning, maintaining a consistent commitment to eugenic policies and programs until the 1960s.[24]

Paradoxes of Paternalistic Compassion

In 1947 Oliver moved to the University of Texas at Austin, and Reed assumed the directorship, which he held until 1975.[25] Following in Oliver's footsteps, Reed stated that institutes such as the Dight were necessary vehicles for the successful pursuit of eugenics in the United States: "In my opinion the most effective way of bringing about Eugenic [sic] improvement in the country is to have a Human Genetics Institute in each state similar to the Heredity Clinic at the University of Michigan or the Dight Institute."[26] Furthermore, Reed was eager to expand on the research that already had been conducted in Minnesota by institutional superintendents and university colleagues on conditions such as Huntington's disease, diabetes mellitus, and, most of all, mental retardation.[27]

In part, Reed was motivated by the ready availability of source material for a research study. In the 1940s, he orchestrated the transfer of a large archive of the Eugenic Record Office files to the Dight Institute, hoping they could be tapped for the development of research projects.[28] One of the largest caches were the records of a study of 549 patients and their families conducted by ERO field-workers at the Faribault State Home (previously called the Minnesota School for the Feebleminded) from 1911 to 1918, which sought to identify feebleminded patients and, fulfilling eugenic prophesy, prove that the unfit beget the unfit across generations in sprawling and unruly families. This project replicated family studies such as *The Jukes* and *The Kallikaks* and, once completed, was published in 1919 as *Dwellers in the Vale of Siddem*. Written by Faribault's superintendent, Arthur C. Rogers, and field-worker Maud A. Merrill, this tract depicted Minnesota's feebleminded as the "ill-nurtured, ill-kempt gutter brats" who carried "on the family traditions of dirt, disease, and degeneracy."[29] They warned of a multitude of defectives on the brink of unchecked reproduction and bad breeding that would harm the state unless curtailed through segregation and sterilization.[30]

Working with his wife Elizabeth, also a PhD, and the Swedish geneticist Dr. Jan A. Böök, Reed soon embarked on a follow-up family study.[31] The Minnesota Human Genetics League and the state's Division of Public Institutions funded the study, supporting Elizabeth Reed for more than 15 years as she tracked the kinships of long-term Faribault patients.[32] After meeting a set of diagnostic and institutional criteria, above all that all subjects register an IQ score of 69 or lower, Reed was able to identify 289 families for inclusion, and up until 1960 she fastidiously collected their health histories. This massive effort resulted in the tome *Mental Retardation: A Family Study*.[33] In contrast to the preceding ERO researchers, the Reeds found that the etiology of mental retardation among the largest single group (123, or 43%) could not be definitely diagnosed, and it was "unknown as to whether genetic or environmental factors were of primary importance."[34] Yet combined, a higher number of 139 patients, or 48 percent, were placed in the primarily genetic category (84, or 29%) and the probably primarily genetic category (55, or 19%). What is striking is that even with inconclusive evidence to support this claim, the Reeds stressed the role of heredity in mental retardation. As they wrote, "The most obvious implication of these studies is that the greatest predisposing factor for the appearance of mental retardation is the presence of retardation in one or more relatives of the person concerned."[35] They were optimistic that organic factors for mental retardation would continue to be discovered, just as the chromosomal anomaly associated with Down syndrome had been identified during the final stages of their research.

Reed began his work on mental retardation during a moment of transition that he and the Dight Institute hastened. In the early twentieth century, people considered as having mental retardation included a wide range of individuals, many classified as feebleminded, some with what we would today recognize as intellectual disabilities, and others so labeled because they came from poor backgrounds or transgressed conventional sexual and gender norms. For several decades, eugenicists succeeded in branding this social group as dangerous, deviant, and destructive to society and the "germplasm" and campaigned for their segregation and compulsory sterilization. By the 1940s, and especially after World War II, parents of children with mental retardation began to challenge these stereotypes.[36]

Critical to this change were the many local organizations that sprouted up across the country starting in the late 1940s and early 1950s. In many instances, assertive parents who would not take no for an answer lobbied state legislatures for programs, started educational and training classes, and raised the visibility of children and people whose bodies and lives had been mistreated and undervalued.[37] The parents acted within a larger network of professionals that included pediatricians, psychiatrists, psychologists, educators, and, increasingly, medical geneticists. For some parents, this meant raising their child at home even temporarily, and for others with the financial resources, it meant placing them in one of the country's few well-run and caring institutions.

Reed's approach to mental retardation encapsulated his conflicted approach to genetic counseling. After publication of *Mental Retardation*, he and his wife thought that their data and findings would help to stimulate voluntary sterilization of people with mental retardation and, moreover, once this practice became "part of the culture of the United States, we should expect a decrease of about 50 percent per generation in the number of retarded persons."[38] Looking ahead, Reed hoped that genetic technologies and reproductive medicine eventually would enable "primarily genetic traits, such as Down's syndrome, galactosemia and phenylketonuria (PKU)," to "be reduced appreciably in the population."[39] He believed that in time more sophisticated medical genetics would fulfill the larger eugenic goal of improving the gene pool. This would become more evident by the late 1960s; against the backdrop of

A couple consulting a medical geneticist at the University of Michigan's Heredity Clinic about the possible occurrence of Down syndrome in their second child. At the time this photo was taken, in 1957, the role of advanced maternal age was recognized as a risk factor but trisomy 21 had not yet been discovered. About the couple: "Not only had they no knowledge of previous Mongolism in either family, but they have been rather proud of the scholastic achievements of their families, and the birth of a Mongolian idiot has shaken them considerably." *Source*: Charles F. Wilkinson Jr., "Heredity Counseling: Genetics Clinics and Their Preventive Medical Implications," *Eugenics Quarterly* 4, no. 4 (1957): 205. Reproduced by permission.

the increasing availability of amniocentesis and hope for the decriminalization of abortion, Reed believed that his well-educated clients would make the rational and right decision to terminate an affected pregnancy. He described the case of one mother who was the carrier of a chromosome 21 translocation who "did not wish to produce several children with Down's syndrome" and was relieved to use amniocentesis. Learning that her embryo was affected, "naturally, she obtained a therapeutic abortion."[40] As he explained, "with the advent of amniocentesis and selective abortion this couple" would produce "normal children only."[41] Given his pessimistic assessment of the impact of children with disabilities on parents and families, Reed counseled parents that the benefits of institutionalization outweighed the burdens of raising a retarded child at home. As he advised the mother of a girl with congenital microph-thalmia, her child would be better trained by the staff at the institution, would not be subjected to insults from normal children, and would not take precious time and attention away from the raising of normal sib-lings.[42]

Yet, even as Reed stigmatized, he planted the seeds for the movement toward disability rights and empowerment. If Reed wanted to stop par-ents from producing children with disabilities, he also wanted to help parents with mentally retarded children cope with the embarrassment and despair associated with having a "defective" child. Reed was aware that he frequently was the first medical professional encountered by par-ents as they sought to understand their child's condition and determine what to do next. As Ohio pediatrician Israel Zwerling explained in 1954, this interaction was pivotal: "Appropriate handling by the physician can turn a potentially devastating experience into the foundation for a satis-factory adjustment to the problem and to the child."[43] One study of the experiences of parents, conducted by a mother involved in the Mary-land Society for Mentally Retarded Children, found that only 25 per-cent of them were satisfied with their interactions with professionals and experts.[44]

Reed's consultations usually began with considerable pushback from parents, as he found many unwilling to accept the truth:

> My experience in general is that the people who are the most resis-
> tant to the idea of heredity probably are the parents of the mentally
> retarded. The social stigma which goes with mental retardation is
> very considerable and they feel that if there is a genetic background
> for mental retardation it classifies them as being part of a Juke or
> Kallikak family. It is a condition which they would deplore; they do
> not want to be part of such a family and do not wish to accept the
> idea, even if they actually belong to a family with many cases of retar-
> dation in it.[45]

Reed handled this resistance well and seems to have won over his clients, in part because he exonerated parents from blame and guilt by invok-ing basic evolutionary principles about genetic variation and mutation. As Reed told the Minnesota Society for the Mentally Retarded in 1952, "The parents are not to blame for heredity received from their ancestors. Usually they are completely unaware of it. Parents are responsible only for the heredity of future generations and not for that of past genera-

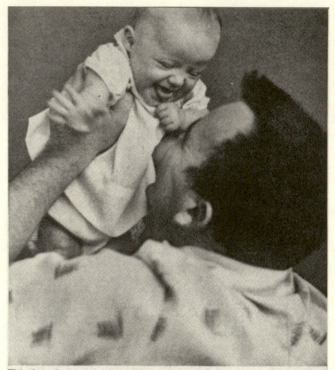

This is a happy baby—and a happy father, too. The period of soul-searching, of self-pity and helplessness is over. Lisa is accepted for what she is and what she has to offer.

A photo that captures the love between a father and his daughter, who has Down syndrome (then called mongolism). This image is the culmination of a story in which a couple is shocked and saddened to learn that they have given birth to a disabled child but soon come to cherish her and decide to reject convention by raising her at home rather than placing her in an institution. This photo narrative, published in *Today's Health* and distributed by the National Association for Retarded Children, illustrates how attitudes toward Down syndrome and intellectual disability began to change in the late 1950s, owing in large part to parental advocacy organizations. *Source*: "They Discovered a New Dimension of Love," *Today's Health* 37, no. 2 (1959): 23. © 1959 American Medical Association. All rights reserved.

tions."[46] Reed's approach attenuated the shame associated with having a child with mental retardation, a psychological process that assuaged many parents.

In the 1950s, parents, including Pearl Buck and Dale Evans Rogers, wrote best-selling memoirs about the joy and worth of their children with mental retardation.[47] This trend touched the national political arena in 1962 when Eunice Shriver published "Hope for Retarded Children" in the *Saturday Evening Post*. Frankly discussing her sister Rosemary, Eunice Shriver wrote, "Like diabetes, deafness, polio or any other misfortune, mental retardation can happen in any family. It *has* happened in the families of the poor and the rich, of governors, senators, Nobel prizewinners, doctors, lawyers, writers, men of genius, presidents of corporations—the President of the United States" (italics in original).[48]

In her memoir *This Is Stevie's Story*, Dorothy Garst Murray chronicled her son's experience. Born in 1945 with what was later diagnosed as both retardation and autism, Stevie was Murray's greatest delight and the impetus for her leading role in the Virginia Association for Retarded Children. Breaking with convention, Murray embraced an "attitude of frankness and openness" about her son's condition: "From the very first day we realized that our little son would probably be a mental cripple, we had absolutely no feeling of shame or disgrace."[49] Murray, like many other parental advocates, rebuffed theories of "good" or "bad blood" being responsible for her son's difficulties. Instead, she highlighted social discrimination against the disabled and the shunning of their families. Murray implored, "Unless we as parents of retarded children make a very conscious effort to rise above being affected too much by what people will think, we may find ourselves floundering in a morass of despair from which it is all but impossible to extricate ourselves."[50] As a parental advocate for her child, Murray exemplified the mothers and fathers who began to turn the tide of opinion about and approaches to people with disabilities in America. Like her contemporaries, however, Murray did not question the logic of prevention and hoped that advances in medical science and genetic research eventually would lead to the end of mental retardation.

Writing in various venues, these authors wanted to free their family members from negative depictions and labels. As disability historian Allison Carey argues, this oppositional effort led many parents to portray their child as an "idealized eternal child," who could fit comfortably in a normal family structure.[51] Reed contributed to the construction of the

At a day camp on her Maryland estate, Eunice Shriver enjoys a spirited ride with gleeful retarded children.

How the Kennedy family's own misfortune spurred the fight against a widely misunderstood affliction.

Hope For Retarded Children

By EUNICE KENNEDY SHRIVER

Eunice Kennedy Shriver was an early champion of the rights of people with disabilities, motivated by the difficult diagnostic and treatment experiences of her sister, Rosemary. Ms. Shriver pushed her brothers, especially John once he became president, to support research and care institutions for the disabled. Her persuasiveness was instrumental to the formation of the President's Panel on Mental Retardation in 1961. *Source*: Eunice Kennedy Shriver, "Hope for Retarded Children," *Saturday Evening Post*, September 22, 1962, 71. "Hope for Retarded Children" article © SEPS licensed by Curtis Licensing, Indianapolis, IN. All rights reserved.

child with mental retardation as innocent, which in turn made it easier for him to deliver his message with much compassion. Reed understood the anxiety and anguish that parents could feel upon receiving a diagnosis of mental retardation. Perhaps in part this was personally motivated, given the Rh incompatibility between Reed and his wife, who had opted for surgical sterilization following the medically dangerous pregnancy of the couples' third child. Explaining that decision, Reed averred, "The geneticist, of all people, should not knowingly take a large chance of producing a defective child. What is good for other families with problems due to heredity is good for the geneticist too."[52]

In consultations, Reed used clear and direct language, seeking to be a scientist who "spoke from his heart."[53] Parents might not have liked the content of the message, but they liked the messenger. As the father of a child with mental retardation who was institutionalized shared with Reed in a letter, "As I think back on our experience, I think we would have been glad if someone before you had told us what we might expect rather than the 'she'll outgrow it' and 'children develop at different rates; don't be concerned' line."[54] Furthermore, Reed's point about genetic exculpation soothed parents of children with mental retardation. Following his advice, many clients, in order to complete their ideal family size, went on to adopt seemingly healthy and normal newborns that were consciously seen as replacements for the usually forgotten child. Reed's files contain numerous cards and notes from parents expressing their deep appreciation for his counsel. In a 1954 letter, one couple thanking Reed for his consultation and information wrote, "It has helped us more than we can say. We know exactly what we are up against, and are trying again with no thought but that if we do happen to produce a normal child, how happy we'll be. If not, we'll use adoption as the method to round out our family. It would be impossible for us to have the matter-of-fact attitude we do now without the help you have given us."[55]

Much more than his counterparts at other American heredity clinics, Reed was closely involved with local and state branches of associations for retarded children. According to parent advocates, as soon as he set up shop at the Dight Institute, Reed "immediately became interested in helping these parents understand the 'why' of mental retardation and the chances of an additional retarded child occurring in a family where there had been one or two."[56] He joined the St. Paul and Minnesota associations for retarded children and regularly spoke at the local and state meetings. Reed earned much praise not just for being a good communicator of the science of heredity but also for "the compassionate man inside who most impresses us. Thank you for your unfailing kindness and patience."[57] In 1963, the St. Paul Association for Retarded Children expressed gratitude for Reed's participation in their fall meeting and for "presenting so many informative and interesting facts about Down's syndrome."[58]

Yet Reed's compassion never left the confines of medical paternalism. There is little evidence that Reed ever modified his viewpoint about the benefits of institutionalizing most children with mental retardation and the need to prevent the future birth of children with disabilities. In correspondence with Elizabeth Boggs, who served on John F. Kennedy's panel on mental retardation, Reed bluntly wrote, "I am perfectly well aware that the parents love the retarded child and give it more than its share of attention, but I can not believe that anyone prefers to have retarded children."[59] And in 1971 he reflected, "I have provided genetic counseling for over 3,000 persons in the last 23 years, and I have never had a couple who would voluntarily produce a child known to have a defect such as Down's syndrome or any other kind of mental retardation."[60] Until the end of his career Reed supported restrictions on "the childbearing of the mentally retarded." In one study, Reed and a colleague "calculated that if no retarded person reproduced, there would be 17 percent fewer retarded individuals in the next generation. Although the net reduction in the number of retarded persons would in fact be modest (from 2 percent to 1.7 percent), it would obviously be a significant 'improvement.'"[61]

By laying the groundwork he did in the 1950s, Reed contributed to the blossoming of parental advocacy for people with mental retardation, as well as the subsequent emergence of intellectual disability activism in the United States. Reed saw a steady stream of clients from the 1940s to the 1970s, responded to hundreds if not thousands of letters of inquiry about potential genetic conditions, frequently gave speeches, taught courses for decades, and published numerous books and articles. Together the Dight Institute and Minnesota Human Genetics League fostered bonds among the parents of children with mental retardation and pushed for the establishment of a state genetics unit in Minnesota. Founded in 1960, this public health entity, which was the first to implement PKU newborn testing in the country, coordinated many parents of mentally retarded children and fortified a growing social network.[62] Through all of these efforts Reed helped to create a framework for classifying and treating mental retardation. His genetic counseling and that of his colleagues—particularly at the Heredity Clinic at the University of Michigan and Wake Forest University—helped to fix the label of mental retardation or Down syndrome and to bring hereditary knowledge to the parents of children with mental retardation. Unlike some physicians who were incommunicative or provided obtuse explanations of a child's condition, Reed was sincere and consolatory.

Contested Labels: From Mongolism to Down Syndrome

The anthropologist Paul Rabinow has written about "biosociality" as a new form of identity making linked to genomic medicine, especially as individuals with rare genetic conditions or at-risk status find each other on the Internet and build new communities in which the point of connection is a diagnosis and medicalized selfhood.[63] Yet biosocial communities, especially those that seek self-empowerment and identity reinforcement, reach back before the era of genomic mapping. As shown, the rise of organizations for disabled groups, many formed by the parents of children with disabilities, was intertwined with the emergence of heredity clinics and genetic counseling. Many parents found each other through their family physicians or medical geneticists like Reed. Building on these proliferating local associations, the National Association for Retarded Children was founded in 1950. Even though Reed and his contemporaries tended to support the institutionalization of children with mental retardation, parents soon turned this convention on its head, advocating for home care and early intervention.[64]

Human geneticists, working in clinical settings, as well as in cytogenetics and biochemical genetics, have helped to till the soil for new biosocial identities through labeling and relabeling. Today, Down syndrome is recognized as the most common genetic disorder associated with mental retardation, usually ranging from mild to moderate, with an estimated prevalence of 9.2 cases per 10,000 live births.[65] Initially, however, Down syndrome was referred to as mongolism, coined in the mid-1800s by the British physician John Langdon Down, who identified the condition as one with distinct facial, cranial, and other physiological features. Siding with the more progressive wing of Victorian evolutionary science, Down mobilized his ideas about the existence of the "Great Mongolian Family" to support the theory of monogenesis, believing that "Mongols" represented an atavistic throwback in the long march of human evolution. Down devised the term "Mongol" or "Mongoloid" in the 1860s as an explicitly racialized construct that conflated low mental status with arrested or stunted evolutionary development.[66] During the early twentieth century, physicians used this term as a catchall for individuals who today likely would be diagnosed along a wide continuum of conditions associated with developmental disabilities, including Down syndrome, cri du chat syndrome, PKU, X-linked mental retardation, and environmental conditions.[67]

People with Down syndrome and their parents were especially sensitive about negative labels and nomenclature. As one mother wrote in the 1950s, "I had heard the word 'Mongoloid' but I really thought it was a monster. I didn't know what a Mongoloid was. And I said, 'Are you going to let me see the baby?' And they did show the baby to me . . . and she looked alright to me."[68] As Wolf Wolfensberger, who spearheaded the movement for the normalization of people with mental retardation, wrote in a coauthored paper in 1974, parents of children with mental retardation chafed at the long list of negative labels, including abnormal, mongoloid, monster, and dumb, attached to their children. One of Wolfenberger's key findings was that even if parents disliked one or two demeaning labels in particular, they abhorred the process of labeling, which had the discursive effect of shrinking their multifaceted child to a singular diagnostic category.[69] When people with intellectual disabilities and their allies moved into activist mode in the 1970s, one of their great refusals was reduction to a label. As the 1978 resolution of the self-advocacy group Project Two, organized by Nebraskan Paul Loomis, proclaimed, "We believe that we are people first, and our handicaps are second. We wish people would recognize this and not give us a tag like 'handicapped' or 'retarded.'"[70] Indeed, this rejection of reductionism sits at the core of the social model of disability, which instead understands disability holistically, as one facet of an individual's complex personality, physiology, and identity.[71]

Jérôme Lejeune's 1959 discovery that "mongolism" was caused by trisomy 21 prompted greater precision in diagnosis and a revised nomenclature. Within one year of this discovery, Curt Stern, professor of genetics at the University of California at Berkeley, launched an informal campaign to banish the term "mongolism" from the genetics lexicon and find a more scientific and neutral term for this syndrome. He sent letters to leading geneticists such as Lionel Penrose, Gordon Allen, Paul Polani, and Raymond Turpin, prompting a statement published in the *Lancet*. As he explained, "mongolism" and "Mongolian idiocy" had "undesirable connotations" and were "not related to the segregation of genes derived of Asians." Stern recommended that "designations which imply a racial aspect of the conditions be abandoned."[72] Hideo Nishimura, professor of anatomy at Kyoto University, provided more culturally nuanced criticism, writing that many Japanese probably believed that this term corresponded to people living in Mongolia, a concept that was "discourteous" to them.[73] Furthermore, Stern maintained that the continued use of this "embarrassing or repugnant" term was insulting to esteemed Chinese and Japanese colleagues.[74]

Stern and the working group proposed various replacement terms, such as Langdon-Down anomaly, congenital acromicria, chromosome 21 syndrome, or trisomy 21 anomaly. After many iterations, Stern obtained the signatures of about twenty leading geneticists, including Lejeune, who initially objected. Irene Uchida at the Children's Hospital of Winnipeg agreed wholeheartedly and recommended to Stern that "reprints of the letter" to the *Lancet* "be made available to the various Associations for Retarded Children in the United States and Canada which play a great role in the education of public attitude towards these unfortunate children."[75]

Even so, "mongolism" continued to appear in medical publications and persisted in the popular lexicon into the 1970s. Nevertheless, Stern's effort constituted an essential stepping stone in the nosological consolidation of Down syndrome as an identity category. By 1979, a new generation of parents had established the Down Syndrome Society of America.[76]

Changing Approaches and Attitudes

Negative stereotypes about Down syndrome and mental retardation did not disappear with a change in nomenclature and arguably did not alter substantively among genetic counselors until the 1980s. In one of the first books on genetic counseling written for a popular audience, Joan Marks, long-standing director of Sarah Lawrence's genetic counseling master's program, and medical journalist David Hendin framed disability squarely within the parameters of disease prevention and used sensational language drawn from the "wars" against cancer and infectious disease to discuss genetic disorders. Parents needed to "protect themselves against genetic disease," families were "victimized" by such conditions, and their children knew only suffering. Describing a case in

"MONGOLISM"

SIR,—It has long been recognised that the terms "mongolian idiocy", "mongolism", "mongoloid", &c., as applied to a specific type of mental deficiency, have misleading connotations. The occurrence of this anomaly among Europeans and their descendants is not related to the segregation of genes derived from Asians; its appearance among members of Asian populations suggests such ambiguous designations as "mongol Mongoloid"; and the increasing participation of Chinese and Japanese investigators in the study of the condition imposes on them the use of an embarrassing term. We urge, therefore, that the expressions which imply a racial aspect of the condition be no longer used.

Some of the undersigned are inclined to replace the term "mongolism" by such designations as "Langdon-Down anomaly", or "Down's syndrome or anomaly" or "congenital acromicria". Several others believe that this is an appropriate time to introduce the term "trisomy 21 anomaly" which would include cases of simple trisomy as well as translocations. It is hoped that agreement on a specific phrase will soon crystallise if once the term "mongolism" has been abandoned.

GORDON ALLEN
Bethesda, Maryland.

C. E. BENDA
Waverly, Massachusetts.

J. A. BÖÖK
Uppsala, Sweden.

C. O. CARTER
London, England.

C. E. FORD
Harwell, England.

E. H. Y. CHU
Oak Ridge, Tennessee.

E. HANHART
Ancona, Switzerland.

GEORGE JERVIS
Letchworth Village, New York.

W. LANGDON-DOWN
Normansfield, England.

J. LEJEUNE
Paris, France.

HIDEO NISHIMURA
Kyoto, Japan.

J. OSTER
Randers, Denmark.

L. S. PENROSE
London, England.

P. E. POLANI
London, England.

EDITH L. POTTER
Chicago, Illinois.

CURT STERN
Berkeley, California.

R. TURPIN
Paris, France.

J. WARKANY
Cincinnati, Ohio.

HERMAN YANNET
Southberry, Connecticut.

A 1961 letter to the *Lancet* recommending that the diagnostic label "mongolism" be replaced by something less offensive and more scientific. This effort was spearheaded by Curt Stern from the University of California at Berkeley and set the stage for the eventual articulation and acceptance of the term "Down syndrome." *Source*: Reprinted from *The Lancet* 277, Gordon Allen et al., "Mongolism," 775, Copyright 1961, with permission from Elsevier.

which a pregnant mother underwent amniocentesis and found out that she was carrying a fetus with trisomy 21, she and her husband deliberated with the genetic counselor and ultimately decided they could not bear the emotional, logistical, and financial drain of raising a child with Down syndrome. According to Hendin and Marks, "the type of preventive genetic counseling" that this couple underwent was "ideal."[77]

Soon after the book's publication in 1978, the mother of a child with mental retardation wrote to Marks to complain about the section on Down syndrome: "I did not find any mention in your book about the positive side of raising a retarded child at home . . . [i]f I might offer some polite criticism. I think it is unfortunate to link Down's syndrome and Tay Sachs disease in that context because Tay Sachs is so much more dismal since it is inevitably fatal. A child with Down's syndrome is a joy." This mother requested that Marks, when she revised her book, "give the other side of the story," which she had found decidedly lacking among "genetic counselors, social workers, pediatricians and other professionals," few if any of whom encouraged home care or discussed promising results from early intervention and infant stimulation programs.[78] Marks indicated in a thoughtful letter that the mother was right to criticize her conflation of Down syndrome and Tay-Sachs disease and that she would consider revising this section. Marks also reiterated her commitment to the psychosocial impulse of genetic counselors to strive "to present to parents, in an unbiased way, information which will enable them to make the decision that is right for them."[79]

By the 1980s, genetic counselors were changing their outlook, in response to activism among those with intellectual disabilities, including children and adults with Down syndrome who spoke with their own voices, told their own stories, and demonstrated that many people with trisomy 21 could function well in society.[80] In an illuminating article in the *Journal of Genetic Counseling*, Campbell Brasington, a genetic counselor in Charlotte, North Carolina, describes the dramatic turn she made in her approach to counseling. When she began her career in the 1980s, representations of children with Down syndrome and mental retardation were uniformly negative. Not only did photographic images usually show an unappealing "abnormal" child with Down syndrome, genetic counselors tended to convey the message that such a child would be nothing more than an emotional, financial, and psychological burden that would drain and disrupt the parents and any existing children in their family. Brasington described her initial stance: "I bought into the medical model of disability that with enough therapy and the right services provided by the 'system,' we could 'fix' the child and make them more 'normal'"; furthermore, she was of the mind that "Down syndrome was a devastating and mostly tragic event."[81] However, once Brasington started interacting with families with children with Down syndrome, she became aware of her biases and decided that she needed to overhaul her methods. She immersed herself in disability-positive literature and ultimately changed how she communicated diagnostic results indicating a pregnancy with Down syndrome. She switched from delivering "bad news" to "unexpected news" that was both realistic and affirming. Nowadays, Brasington counsels that families "who have a child with Down syndrome are typically no different from other families. Children often attend school in an inclusive classroom and participate in a variety of activities in the community with their typically developing peers. They are more like other children than different."[82]

By the 1990s it would have been nearly impossible to find a genetic counselor who would have adhered to Oliver's dictum to paint as negative a picture as possible of intellectual disability. As discussed in chapter 1, in 1992 Judith Tsipis established a genetic counseling program at Brandeis University that was explicitly informed by the disability studies critique of the biomedical model of disability.[83] Annette Kennedy, a psychologist trained at the Massachusetts School of Professional Psychology, played an instrumental role in designing the psychosocial portions of the curriculum at Brandeis. "Counseling Theory and Technique," a course that she taught from 1993 to 2000, highlighted writings by and for people with disabilities. Kennedy ensured that students interacted in meaningful ways with adults with disabilities, an experience that almost always prompted them "to start to think differently, broadening their sense of what was possible," particularly with regard to continuing, rather than terminating, a pregnancy.[84] Furthermore, Tsipis has found a variety of ways to incorporate the disability critique into the curriculum, including readings and guest speakers. Through community placements in schools or clinics serving children or young adults with disabilities, Brandeis students acquire "hands-on" experience and become familiar with the multitude of existing programs and entitlements. Finally, Tsipis and one of her faculty members, Barbara Lerner, conceptualized and implemented the Family Pals Program, which pairs each student with a family that includes a child with a disability so that students learn "outside a clinical setting so that they will be able to provide sensitive, balanced and knowledgeable counseling to the families they will be serving in the future."[85]

Genetic counseling also became more disability-friendly because of the orientation and experiences of the master's-degreed genetic counselors who took positions around the country. For example, Wendy Uhlmann was drawn to the field because of experiences with her late father, who had multiple sclerosis (MS) for almost 50 years and was a wheelchair user. As she explains, "My father taught me the important lesson about living life to the fullest with a disability rather than having the disability define the life that he lived."[86] She observed how he took ownership of his health care, actively seeking out information about treatment options and willing to "fire" physicians who did not treat him as a partner. In the 1970s and early 1980s, her father frequently wrote letters, published in the *Washington Post*, advocating for handicapped parking spaces, curb cuts, and wheelchair access.

Uhlmann was keenly aware of the challenges and social isolation that her father and family faced because of his MS and the lack of open discussion about his diagnosis or connections to other families living with MS. Indeed, it was not until she got to Oberlin College to start her undergraduate education that Uhlmann met another student whose parent also had MS. Once Uhlmann learned about genetic counseling as a career option, she was attracted to the field because of its emphasis on providing information and empowering patients to be active participants and make informed decisions about their health care. In addition, she saw how genetic counselors could help connect patients to support groups and other critical resources. Currently Uhlmann strives to provide such services as a genetic counselor and coordinator of the University of Michigan's Medical Genetics Clinic.[87]

Robin Bennett, genetic counselor at the University of Washington's Medical Genetics Clinic, also links her interest in genetic counseling to the powerful impressions made by the son of a close family friend. Growing up in Mercer Island, Washington, Bennett interacted with this boy, who had severe intellectual disabilities. At the same time, she learned from the extensive family genealogies assembled by her maternal grandmother about an ancestor who lived in the 1800s most likely with Down syndrome, requiring a great deal of medical care and supervision. Combining these two experiences inspired Bennett "from an early age to want to help people with birth defects."[88] After an undergraduate degree at the University of Washington, Bennett completed her master's in genetic counseling in 1984 at Sarah Lawrence College. Bennett was happy to receive a job offer from the University of Washington, where she was hired as one of the first master's-level genetic counselors at the Genetics Clinic that had been started by Arno Motulsky in the 1950s.

Unlike many genetic counselors who work at least for a short period in prenatal diagnosis, Bennett has worked entirely with adults and adolescents with genetic conditions. In this capacity, Bennett ended up providing solace for the mother of the boy with severe disabilities whom she had known as a child. Once back in Seattle, Bennett arranged for a geneticist colleague to visit him (now a grown man residing with his caregiving sister), who soon diagnosed him with Angelman syndrome, a revelation that allowed his mother to understand that his condition was not the result of her actions or behavior during pregnancy.[89] Bennett believes that these kinds of moving interactions with family and friends have enabled her to "hear what is in people's hearts, and reflect back what they need to hear."[90]

The founder of the genetic counseling program at the University of Wisconsin at Madison, Joan Burns, was attracted to the expanding field because of both her strengths in the biological sciences and the fact that her third child was born with severe developmental disabilities.[91] After having received a master's degree in zoology and genetics, Burns's interest in disability from scientific and personal perspectives prompted her to earn another master's degree in social work in the 1970s. She explains that she wanted to know more about "mental retardation and began wondering how families found out about recurrence risks. I became involved in the Association for Retarded Children (as it was called at that time), and came in contact with social work professionals who were helping families deal with the many aspects of living with a child with disabilities."[92] Guided by these concerns and with Burns at the helm, Madison's program accepted its first cohort of students in 1976.

For Katy Downs, a genetic counselor who has been involved in research projects at the University of Michigan, her pathway to genetic counseling started in the disability community. In the 1970s Downs was a sign language interpreter in the San Francisco Bay Area and very active in the deaf community. After seeing the need for sensitive genetic counseling for deaf adults, Downs enrolled in the genetic counseling program at the University of California at Berkeley. Directed by Seymour Kessler, this program was part of a new initiative in community and integrative health. In keeping with this mission, and given the centrality of the East

Bay to the disability and independent living movements, Downs spent her first year not in medical clinics but interning with the Association for Retarded Citizens. She worked with the Center for Independent Living while in school and as an educational genetic counselor at Gallaudet University after leaving California. Arriving to the field from the disability community, Downs has rejected the "medical-pathological model" and does not view deafness as a disability.[93] Acknowledging the "conflict of cultures between the Deaf community and genetics," Downs has worked to make genetic counseling incorporate cultural considerations and become more attuned to diversity in terms of race, ethnicity, class, sexual orientation, language, and disability.[94]

To a great extent, the past several decades have seen the broader social acceptance of people with disabilities envisaged by Cooke in the 1960s. Starting in the 1970s, a series of laws banning discrimination against people with disabilities, guaranteeing access to transportation, and mandating other services in health and education have been passed, culminating in the 1990 Americans with Disabilities Act. These legal strides have been accompanied by the increasing visibility of disability rights on the grassroots level and disability studies in the academic realm.

Search for Consensus

When it began in the 1940s, genetic counseling was built around a code of prevention. For instance, throughout his career Reed believed that clients making rational decisions about their reproductive future would weigh risks rationally to avoid producing mentally retarded children. With the emergence of prenatal diagnosis in the 1970s, fetuses that tested positive for Down syndrome and other genetic and chromosomal disorders could be aborted therapeutically. As the disability rights movement gained ground, so too did new reproductive and genetic tests and technologies, allowing an expanding group of mothers and parents to terminate pregnancies in order to prevent mental retardation for the emotional, psychological, and economic sake of the family and society.[95]

Over the past several decades, the prevention rationale has come under sustained criticism, most forcefully by disability scholars and activists, who express frustration at the slow acceptance of the social model of disability among some genetic counselors. Erik Parens and Adrienne Asch, for example, contend that genetic testing presumes that a multifaceted human being can be reduced to a disease or genetic marker.[96] Even if eugenic rationales are not an explicit component of prenatal screening and testing programs today, the cost-benefit and preventive arguments continue to be common refrains.[97] Some disability scholars, such as Tom Shakespeare, have sought to find a balance between the medical model and the social model, which can negate important physiological or biological aspects of disability that some people might quite reasonably believe are debilitating or harrowing enough to validate the termination of an affected fetus. Shakespeare suggests separating impairment, which limits an individual's physical and mental abilities, from disability, which must be understood in a broader social, economic, and legal matrix of structures and experiences. Shakespeare argues that doing so can help people with and without disabilities more sensitively utilize prenatal screening to make difficult decisions about proceeding with or ending a pregnancy after the detection of a genetic disorder or chromosomal anomaly.[98] Echoing contemporary genetic counselors, Shakespeare advocates for the production of "good quality, balanced information about Down syndrome, spina bifida and other conditions detectable through ultrasound scan, chorion villus sample or amniocentesis."[99]

Some geneticists, most notably Linda and Edward McCabe of the University of Colorado Denver, have criticized the subtle and not so subtle forces placed on women both to undergo prenatal screening and, if given a diagnosis of trisomy 21, to terminate the pregnancy, identifying patterns of coercion and discrimination in terms of Down syndrome information delivery. The McCabes write, "As medical genetics professionals, each of us must ask whether we have been influenced by the social perspective that devalues the life of the individual with Down syndrome. If we wish to be supportive of families in their choice and provide optimal care to their children, then we must recognize the risk of discrimination and coercion."[100] Notably, two important studies have demonstrated that parents, particularly mothers who have children with Down syndrome or elected to continue pregnancy after receiving a fetal diagnosis of Down syndrome, are dissatisfied with the care, information, and support they have received pre- and postnatally.[101] Legal scholars, examining the limits of informed consent, have echoed these concerns, worrying that not infrequently women are pressured into prenatal testing and then given incomplete or biased information, which does a disservice to them as decision makers and to people with disabilities and their communities.[102]

The large-scale longitudinal studies conducted over the past four decades that calculate termination rates of fetuses diagnosed with a genetic condition or chromosomal anomaly demonstrate that the overwhelming majority of women who undergo amniocentesis and then subsequently receive a positive diagnosis elect to end their pregnancies.[103] This number is the highest, with averages ranging from 81 to 96 percent, for the termination of fetuses diagnosed with Down syndrome (trisomy 21) during the first and second trimesters via either amniocentesis or chorionic villus sampling. It is significantly lower, ranging from 40 to 60 percent, for fetuses with sex chromosome anomalies, such as Turner or Klinefelter syndromes, which parents apparently feel are more socially and economically manageable. Furthermore, several studies have demonstrated that African American and Latina mothers terminate fetuses diagnosed with trisomies at a lower rate than whites, suggesting that class and education levels play an important role in the decision-making process.[104] For example, an analysis of 833 patients with fetuses with aneuploidy (one or more extra or missing chromosomes) at the University of California at San Francisco seen from 1983 to 2003 found that African Americans terminated affected pregnancies at a rate of 76 percent, Latinos at 69 percent, and whites at 84 percent.[105] Other research, conducted in the Atlanta metropolitan area and with larger catchment areas in the Southeast, has found significantly lower rates among African American mothers.[106]

Given these figures, it is not surprising that people with disabilities and their parents have expressed wariness toward prenatal genetic screening, even as genetic counselors today seek to be disability-friendly. In 2007, the American College of Obstetricians and Gynecologists (ACOG) revised its guidelines to recommend that *all* women—not just those over 35—receive prenatal genetic screening, ideally triple or quadruple screening (maternal serum alpha feto-protein, human chorionic gonadotropin, and unconjugated estriol) and nuchal translucency measurement.[107] The Down syndrome community, most notably the National Down Syndrome Society and the National Down Syndrome Congress, were alarmed at this revised recommendation, perceiving it as an attack on their very existence and an attempt to encourage the termination of Down syndrome pregnancies in all expectant women regardless of age or other criteria.

Parents of children with Down syndrome chafed at the revised guidelines as well. In a 2007 *Washington Post* editorial, one mother worried that "these recommendations will have the effect of accelerating a weeding out of fetuses with Down syndrome" and then expressed her "fervent hope: that calls for universal prenatal screening will be joined by an equally strong call for providing comprehensive information to prospective parents, not just about the tests but also about the rich and rewarding lives that are possible with disabilities."[108] Another vocal opponent, the well-known conservative *Newsweek* columnist George Will, charged that with its new recommendations, ACOG wanted to eradicate "from America almost all of a category of citizens, a category that includes Jon," his son, a "sweet-tempered" man in his 30s able to live independently and work part-time for the Washington Capitals, his favorite hockey team.[109]

In response to such criticism, ACOG stated that they simply wanted to ensure that high-sensitivity and noninvasive technologies of prenatal screening that could be performed in the first trimester were universally available. Ideally, such screening—for example, nuchal translucency— would help women with higher degrees of risk to decide whether or not to proceed with the more invasive diagnostic procedures of amniocentesis or CVS. In what appears to have been a successful effort to reconcile these misperceptions and demonstrate the critical role of genetic counselors in communicating the results of prenatal testing in a balanced and empathetic manner, Janice Edwards, director of the University of South Carolina's genetic counseling program, worked with key constituents to organize a two-day conversation around these issues. Ultimately this rapprochement culminated in the drafting of a Concurrence Statement that clarifies the position of obstetricians, gynecologists, and genetic counselors as maintaining no vested interest in "reducing the Down syndrome population."[110] Edwards's leadership in facilitating this conversation and coming to a shared understanding of prenatal screening among physicians, genetic counselors, and the Down syndrome community constituted one important step in forging common ground between genetic counseling and disability rights.

More recently, in 2011, the National Society of Genetic Counselors issued a new position statement on disability, confirming "the goal of the genetic counseling profession to advocate for all individuals and families according to their unique physical, medical, cultural, educational, and psychosocial needs. The NSGC believes that no person should be discriminated against because he or she has a disability."[111] A growing number of genetic counselors have become involved in finding ways to resolve the juxtaposition that "genetic counselors simultaneously represent and advocate for the rights and opportunities of those affected by disabling conditions on the one hand; and offer parents reproductive opportunities to avoid having children with disabilities on the other."[112]

Finding this balance might require that genetic counselors examine their own potential biases toward prenatal testing and the enduring tendency to present a "narrow description of Down syndrome" that focuses "on negative aspects of the condition."[113] Genetic counselors committed to comprehensively changing the field's approach to Down syndrome suggest cultivating much greater interaction between genetic counselors and the disability community on the level of genetic counseling practice, student training, and professional organizations, as well as consistent and probing assessment of the field and its underlying values.[114]

CHAPTER 5

Women

Transforming Genetic Counseling

In 1970 the *New York Times* interviewed Melissa Richter about the innovative program she had launched the previous year at Sarah Lawrence College in Bronxville, New York. A laboratory psychologist who was committed to improving science literacy and expanding career opportunities for returning students, Richter was the visionary behind the world's first graduate program in genetic counseling. Keenly aware of accelerating developments in genetic and reproductive medicine, Richter foresaw the potential for an original health profession that would appeal to smart and personable women who wanted to pursue an advanced degree with practical applications. Initially Richter had devoted her research to rat biology and behavior. However, she recognized that advances in human genetics were transforming knowledge about disease and the delivery of clinical care, especially in the realm of reproductive health. Richter also realized that discoveries in human genetics had the capacity to provoke deep introspection about a person's core identity, life purpose, and legacy. As she told the *New York Times*, "When a genetic problem hits, it hits at the very gut of people, at the questions of what am I and what do I leave to the world. It requires so much intelligence to be able to deal with that."[1]

Forty years after Richter embarked on an unprecedented educational experiment at Sarah Lawrence College, genetic counseling has become a dynamic profession with more than 3,000 certified practitioners. It possesses the hallmarks of any bona fide profession, including a credentialing specialty body, the American Board of Genetic Counseling, and a professional organization, the National Society of Genetic Counselors.[2] Genetic counselors are employed throughout the health care system, and many medical providers rely on the profession's guidelines to communicate sensitive genetic information to individuals and families. From today's vantage point, this appears to be a fairly routine story of the development of a new group of health care professionals with a distinctive portfolio of skills and responsibilities.

However, the birth of the contemporary genetic counselor was neither predictable nor smooth. Instead, this is a story of educational experimentation, audacious expectations, and fortuitous timing that unfolded in parallel with dramatic shifts in American medicine and society. The making of the modern genetic counselor occurred during a period of extraordinary convergence that included second-wave feminism, the decriminalization of abortion, discoveries in medical genetics, the growth of prenatal services and genetic testing, and changing attitudes toward the physician-patient relationship. The principal actors in this story are women, mainly white and middle-class, who were instrumental to the establishment of genetic counseling programs and organizations, and who rather quickly came to dominate the profession. Men certainly participated in the development of master's-degreed programs in genetic counseling, perhaps most noticeably by expanding the pediatric, prenatal, and adult genetic clinics that needed trained counselors. Yet men have consistently made up only a small percentage, in the range of 5 to 10 percent, of the student body in genetic counseling programs. Some male genetic counselors have had a significant impact on the profession in their day-to-day work with clients or in leadership roles at the NSGC. Indeed, as a gendered minority, male genetic counselors have been pioneers in their own right. Nevertheless, the history of the genetic counselor is, to use the alternative word coined by feminist historian Robin Morgan in her classic 1970 anthology *Sisterhood Is Powerful*, much more of a herstory whose key protagonists were and continue to be women.[3]

"Sarah Lawrence's Special Baby": A Child of the Sixties

The late 1960s was a tumultuous decade in the United States and around the world.[4] The year 1968 saw the assassinations of Martin Luther King Jr. and Robert F. Kennedy, the showdown between antiwar demonstrators and police at the Democratic Convention in Chicago, and rising opposition to the Vietnam War. The countercultural Woodstock Festival, the American Indian takeover of Alcatraz prison, and the Stonewall rebellion for gay liberation in New York City all occurred in 1969. Despite its suburban insulation in Bronxville (Westchester County), New York, Sarah Lawrence College was not immune to the unrest that rocked the country in the sixties, especially on college campuses. In March 1969,

Sarah Lawrence students occupied the main administration building for 10 consecutive days to protest tuition increases and demand increased male enrollment (coeducation had begun just one year earlier), racial and socioeconomic diversification of the student body and faculty, and more community involvement in local service programs.[5]

It was during these turbulent months that Melissa Richter was laying the groundwork for Sarah Lawrence's genetic counseling program. The previous year she had begun to envision what she called "Sarah Lawrence's special baby" as one prong of a larger venture that included tracks for elementary health education and psychological training and counseling.[6] Although she never met Sheldon Reed in person, Richter told him in a letter that it was after reading his *Counseling in Medical Genetics* that she began to slant her "course in Evolution and Genetics more and more toward an understanding of genetics in social medicine, and ultimately began developing a proposal for training genetic counselors" in a master's degree program.[7] At the time Richter was Dean of Graduate Studies and being groomed to direct Sarah Lawrence's Center for Continuing Education. Created in 1962, this center aimed to encourage women who had abandoned college education for reasons of marriage, motherhood, or work to return to complete their bachelor's degree or to obtain a professional degree.[8] The center's founder, Esther Raushenbush, Sarah Lawrence's president from 1965 to 1969, described it as a "revolutionary step in the education of women" and viewed it as one component of a concerted effort to bring "equality of opportunity" to women and racial minorities.[9]

Richter saw the Center for Continuing Education as the ideal home for a genetic counseling program, believing that the program would appeal to the center's core constituency, married women in their thirties with two to four children living at home. An adherent of difference feminism, or the idea that women deserved equality and civil rights because they differed in substantial and valuable ways from men, Richter imagined genetic counseling as suited distinctly for women "because they generally are more concerned with health and the preservation of life."[10] A divorcee with no children, Richter was well poised to develop new career pathways for mature women whose lives were in professional and personal transition.

From her corner in Bronxville, Richter was able to take advantage of the spaces opened by the feminist and civil rights movements to attract a widening female workforce, specifically the segment composed of "a new breed of middle class, at least partly college-educated" women.[11] Clearly the moment was ripe for Richter's initiative. Not only would the profession intrigue women interested in health, psychology, and social work, but an expanding menu of genetic tests for metabolic conditions and prenatal diagnostic technologies was changing obstetric, pediatric, and subspecialty care. Richter's recognition of these advances, as well as their implications for patients, prompted her to design the genetic counseling program. Moreover, genetic information now could be weighed in decisions about whether or not to legally terminate a pregnancy. In 1970, three years before the U.S. Supreme Court decision *Roe v. Wade*, New York decriminalized abortion, permitting women to obtain abortions pursuant on informed consent and as long as performed no later than the 24th week after conception.[12] Thus, Sarah Lawrence's inaugural cohort of genetic counseling students started their clinical training as abortion became feasible for patients, and at a time when the feminist health and reproductive rights movement had augmented acceptance of and access to birth control for many women.[13]

Even before the program was announced officially, a handful of enthusiastic women contacted Sarah Lawrence to inquire about application. Through publicity at a conference on genetic counseling sponsored by the March of Dimes and the Medical Society of the County of New York's Special Committee on Infant Mortality, word of mouth in and around New York City and Westchester County, and passing reference in the *New York Times Sunday Magazine*, a sympathetic audience learned about Richter's plans during spring 1969.[14] By the fall of that year, one dozen women had contacted Richter by mail and phone about enrollment.[15] The interest was so great that Richter launched the program one year ahead of schedule, welcoming 10 students in fall 1969 and two more the following semester.[16]

The Perceptive and Persistent Melissa Richter

Born in 1920 in Mount Vernon, New York, and raised mostly in White Plains, Richter was a Sarah Lawrence alumna who earned a PhD in psychology from the University of Connecticut in 1959.[17] After obtaining her degree, Richter taught physiology and anatomy at Vassar College for several years. She joined the Sarah Lawrence faculty in 1963, where she held key administrative positions, including dean of graduate studies.[18] From 1960 to 1968 she also served as principal investigator for a series of research grants from the United States Public Health Service and published articles on induced thirst and calcium deprivation in laboratory rats.[19] According to her colleague and close friend Jacquelyn Mattfeld, then dean of the college, Richter had a profound interest in human and animal variation from the standpoints of both behavioral psychology and evolutionary biology.[20] She was also a vocal advocate of women's education who was eager to expand career options for returning female students, particularly in the health sciences.

Richter's approach to genetic counseling was prescient and original.

A vivacious Melissa Richter in her office at Sarah Lawrence College, circa 1970. *Source*: Jacquelyn Mattfeld. By permission.

She foresaw a budding niche for specialists trained in laboratory, statistical, and psychological aspects of human genetics. As she told the *Sarah Lawrence Alumnae Magazine* in 1971, "Researchers are making new breakthroughs in genetics all the time, discovering new diagnostic tests, new tools to help," but she lamented that "there is nobody to pass these services on to patients. There is a tremendous gap at this point between knowledge and service."[21] Richter's objective was to bridge that gap by training smart and caring women who could communicate effectively with patients, many of whom would consult genetics clinics as prospective or concerned mothers.

In addition to being perceptive, Richter was bold and persistent. Although she had few if any connections to medical geneticists, Richter sent letters to genetics units in universities, hospitals, and state health departments across the country in order to assess the need for a master's program in genetic counseling. In return, she received scattered praise and much criticism. One physician told her, "I am very fearful that the results of the program you are outlining will be a disaster"; another warned her to "stay away from this area"; and yet another exclaimed that her plan was "nonsensical" and a "totally unrealistic approach."[22]

Again and again her respondents, in what can be understood as a kind of paternalistic and professional gatekeeping, emphatically stated that only PhDs or MDs could legitimately provide genetic counseling. Almost all the experts, mainly medical geneticists and physicians, contacted by Richter viewed genetic counselors as mere appendages—aides, helpers, assistants, or, at most, associates—charged with finishing the low-priority duties for which busy physicians had no time or inclination. In what appears to be the only epistolary exchange between Richter and Reed, the man who coined "genetic counseling" responded to the female pioneer by stating that the counselor "must be a competent geneticist," with at least "a master's degree in genetics," and ideally a PhD.[23] Even if encouraging, nearly all the letters expressed scant faith in the capacity of Sarah Lawrence, a small liberal arts college, to adequately institute such a program. Many people might have given up after encountering such resistance, but Richter was buoyed, not disheartened, by the responses, negative and positive, that she received. Frequently, Richter replied to scathing letters with a cheerful thank you and update on how the program was proceeding.[24]

Given Richter's research interests in the social behavior of animal populations enduring stress, it is not surprising that she framed the need

"If we were informed, if we projected the consequences of unrestricted reproduction and we can do this, then we could ... decide whether we want to control our population and if so how this control can take place."

Melissa Richter discusses her motivations for establishing the genetic counseling program in the *Sarah Lawrence Alumnae Magazine* in 1968. Once the program was running, her initial goals of the regulation of reproduction and population control receded, eventually supplanted by a focus on the individual patient and her or his medical, social, and personal circumstances. *Source*: Melissa Richter, "The Effects of Over-Population on Behavior: The Biologist's View," *Sarah Lawrence Alumnae Magazine*, Spring 1968, 12. Sarah Lawrence College Archives.

for genetic counseling in terms of population management and optimization of a group of biological organisms. Notably, Richter delineated the key reasons for the program's creation as "the increase of illnesses caused by inherited diseases; the sizeable proportion of our population suffering from mental or emotional disturbances; and the problems created by over-population."[25] Drafting her proposal the same year that Paul Erhlich published the best-selling and apocalyptic *The Population Bomb*, Richter expressed both stridency and naiveté about the possibilities of regulating the "load of genetic disease that is accruing in our population."[26] Exhibiting faith in the curative power of modern medicine, extrapolated from the wonders of antibiotics and vaccines and a few apparent triumphs in the treatment of metabolic disorders, Richter viewed genetic counseling as a means for reducing hereditary diseases and persuading individuals to make rational decisions about reproduction for their benefit and that of the population at large.

Richter couched her pilot proposal in the neo-Malthusian language of population control. She expressed unambiguous ideas about appropriate and inappropriate breeding, normality and defectiveness, as well as optimism about the prospects of genetic control. In this sense, Richter followed in the footsteps of a preceding generation of medical geneticists such as Sheldon Reed, Nash Herndon, and Lee Raymond Dice who had occupied the inchoate domain of genetic counseling from the 1940s to the 1960s with one foot in the eugenics camp.[27] Ultimately, however, Richter's position was characterized more by metamorphosis than continuity. She initiated Sarah Lawrence's program as the eugenics era of the early to mid-twentieth century was in eclipse. Although she initially accepted eugenic reasoning and never abandoned the idea of prevention, ultimately Richter supported a brand of genetic counseling that emphasized private decision making, reproductive choice, and budding concepts of bioethics. In large part, this reflected Richter's interest in patient care, an area in which she had virtually no experience but much insight. For example, in a lecture before the New England Association of Nurses at Boston College in May 1970, Richter stated that one goal of medical genetics was prevention, such that "everyone can be born free of birth defects." Nevertheless, the paramount aim of the genetic counselor should be improved and dedicated "patient care, regardless of its evolutionary effects," which she touted as the "the primary driving force for all of us who are contributing, or who wish to contribute to this field."[28] As her program moved from the page to the clinic, Richter's focus shifted accordingly, away from philosophical declarations about the regulation of reproduction and population to the logistics of fieldwork placements and the challenges of conveying complex genetic information to patients and clients. The shift of emphasis from population to patient that Richter experienced in a few compressed years soon congealed into a cornerstone of modern genetic counseling in the form of client-centered care.

Location, Location, Location: Bronxville to Manhattan, 15 Miles

Richter was a maverick, but she did not persevere in a vacuum. She certainly would not have been able to get the program off the ground without the backing of Sarah Lawrence's administration. Founded in 1926 as a progressive women's college, Sarah Lawrence possessed a mixture of affluent students and an emphasis on individualized education, supple curriculum, and deep commitment to service learning.[29] Since the 1930s, fieldwork in community sites had been central to the college experience of Sarah Lawrence students. The genetic counseling program resonated with that creative tradition of experimentation. Furthermore, the faculty operated in a fairly unstructured environment that granted considerable leeway in educational design.[30] Rather than rein Richter in, her superiors nurtured and championed her ideas. Richter's most important collaborator and confidante was Jacquelyn Mattfeld, with whom Richter regularly brainstormed about new directions for women's education in science and medicine.[31] Mattfeld, a savvy administrator who later became a provost at Brown University, repeatedly stressed Richter's notable academic talents and personal style to colleagues and potential grantors alike.[32] For instance, in a grant request to the Commonwealth Fund, Mattfeld heralded Richter as "one of those rare persons who combine originality with an uncommon gift for teaching, and genuine administrative ability."[33] Richter also benefited considerably from her warm relationship with Raushenbush, who was pivotal to the development and financial stability of the program. Thanks in large part to the efforts of Mattfeld and Raushenbush, the program started with a $20,000 grant from the Babcock Foundation. Richter built on this initial investment and established a funding pattern that the program's subsequent director, Joan Marks, a psychiatric social worker and Sarah Lawrence alumna, emulated with tremendous success. For example, the program was awarded a multiyear grant from the National Institutes of Health Manpower Training Division in 1970 and a three-year student training fellowship from the March of Dimes in 1974.[34]

The college's proximity to New York City, an epicenter of medicine with many prominent teaching hospitals and, most importantly, recently established genetics clinics, was integral to the program's achievements. It is difficult to imagine any other place in the United States that could have allowed for such extensive and varied internship experiences for genetic counseling students. Richter astutely took advantage of Sarah Lawrence's strategic location. During the three years that she directed the program, Richter built productive relationships with clinical geneticists and allied physicians throughout New York City. Once convinced of the viability of Richter's plan, this network of supportive physicians became instrumental to the program's consolidation and increasing visibility to the medical genetics community at large.

One of Richter's closest New York City partners was Jessica Davis, a pediatrician at Albert Einstein College of Medicine Hospital who had become interested in genetics during her medical training at Columbia University.[35] Davis, also an alumna of an elite women's college (Wellesley) and a tireless promoter of female scientists, was a natural ally for Richter. Indeed, Davis agreed, despite a more than full-time clinical and research load, to codirect the program with Marks during 1972–1973 when Richter took a sabbatical to pursue research on sex differentiation and, unbeknownst to many, battle a recurrence of breast cancer.[36]

In spring 1969, Richter contacted Davis for advice on how to get Sarah Lawrence's program off the ground. Soon thereafter she went to Albert Einstein College to meet with Davis and several of her colleagues. Not only was Davis "quite enthusiastic with her presentation," but she was taken by Richter's persona and vision.[37] Like Richter, Davis believed that an exciting opening existed for master's-level genetic counselors. Based on her clinical work in pediatric genetics, Davis suggested that Richter model the genetic counselor on the medical social worker, an idea discussed in great depth by Reed in the 1950s.[38] Unlike Reed, however, who thought strongly that only MDs and PhDs could adequately do what he called "a kind of genetic social work," Davis and Richter were convinced that the master's-level genetic counselor could combine the strengths of social work with rigorous training in the life sciences to distinguish herself as an independent health care provider. Kurt Hirschhorn, a pediatrician and research geneticist at Mt. Sinai who devoted a great deal of time and energy to helping Richter launch the program, recommended Lynn Godmilow, a social worker who had enhanced his clinical practice and patient care, as a guiding example.[39]

Davis was very helpful to Richter during this formative period; she helped her build "a bridge to the genetics community, which was small and very inward" and not necessarily eager for educational innovation.[40] Reiterating a suggestion made by Mattfeld, Davis urged Richter to identify advisors that could provide feedback and help the program insinuate itself into an enlarging network of medical genetics clinics and services. Sarah Lawrence's first trio of advisors—Hirschhorn, Arthur Robinson of the University of Colorado Health Sciences campus, and John Littlefield, then affiliated with Massachusetts General Hospital in Boston— offered sage counsel and conferred scientific and clinical gravitas to the program.[41]

That proximity to New York City was central to the evolution of Sarah Lawrence's program is illustrated by reviewing field placement sites from 1969 to 1972. Half were clinics in Manhattan that spearheaded the use of new procedures including amniocentesis and maintained state-of-the-art laboratories where enzyme analyses of diseases such as Tay-Sachs could be performed. They included Mt. Sinai Hospital, New York Cornell Medical Center, the Albert Einstein College of Medicine Hospital, Beth Israel Hospital, and the New York State Psychiatric Institute. Outside the city limits, Richter established internships at two state institutions—Creedmoor State Hospital and Letchworth Village—and at the Westchester County Community Mental Health Board. At Creedmoor, Richter received a warm welcome from the director of psychiatric research, John Whittier, who let Sarah Lawrence's students participate in consultations with and observe the clinical care of persons with Huntington's disease.[42] Richter even set up two out-of-state sites, at Massachusetts General Hospital and Chicago Children's Memorial Hospital, which offered summer placements owing to the enthusiasm of, respectively, Littlefield and Henry Nadler.[43]

Manhattan was also where Sarah Lawrence students received training in clinical genetics. When the initial group of 10 women joined the Sarah Lawrence program in fall 1969, they attended most classes, including Mendelian and molecular genetics and social psychiatry, at Sarah Lawrence, but they studied medical genetics at Mt. Sinai and at the Albert Einstein College of Medicine Hospital with Harold Nitowsky and Jessica Davis.[44] In addition, an arrangement with Hirschhorn allowed students to participate in medical conferences at Mt. Sinai Hospital.[45] Furthermore, many in the program's initial graduating classes so impressed their mentors that they secured jobs where they had interned or were recommended for positions at genetics clinics in the New York area.[46]

Although she directed Sarah Lawrence's program for just three brief years before she succumbed to breast cancer, Richter left a lasting impression on colleagues and students alike. Davis remembers that in her soft-spoken and gentle manner, Richter energetically pursued her plan, winning many converts along the way. For example, through her persuasive personality and steadfast conviction, Richter brought the distinguished medical geneticist Arno Motulsky, who at first glance was quite critical of her plan, to her side.[47] Mattfeld remembers Richter as "an amazing teacher, adored by her students, and absolutely creative in the classroom," who enjoyed the rare position of being liked universally by her colleagues at Sarah Lawrence.[48] Looking back at her experience as a member of Sarah Lawrence's first cohort, Audrey Heimler exclaims that Richter "lit a fire under us," as she inspired students about the exciting future that awaited them in clinical genetics.[49]

After Richter died in November 1974, the program thrived under the leadership of Marks, who served as director until 1998. During her 25 years of leadership, Marks amplified the psychosocial component of the

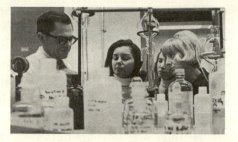

Harold Nitowsky teaching Sarah Lawrence genetic counseling students at
the Albert Einstein College of Medicine
Hospital, circa 1970. These clinical rotations were formative training experiences for budding genetic counselors
and often served as the stepping stone
for a job after graduation. *Source*: Human
Genetics Graduate Program Brochure, circa
1970. Sarah Lawrence College Archives.

program, developed relationships with new clinical sites such as North
Shore University Hospital and Montefiore Hospital, became a prominent figure in the medical genetics community, and was pivotal to genetic counseling's rising national prominence.[50] The American Society
of Human Genetics recognized her contributions in 2003 when Marks
received the Award for Excellence in Human Genetics Education.[51] In
addition, during her tenure as director, Marks founded the Graduate
Program in Health Advocacy at Sarah Lawrence to train professionals to
work in complex health care settings.[52]

New Professionals, New Programs

One of the reasons why Sarah Lawrence's program was able to prosper
was that its high-quality students fit, and helped to carve, an emerging
niche very well. In Richter's words, the students were "a most unusual
bunch," "all highly motivated, deeply concerned to serve and highly intelligent."[53] In short order, they acquired skills that were in great demand
as genetic services became a routine facet of medical care. Nevertheless,
reflecting Richter's interest in the returning students affiliated with the
Center for Continuing Education, the cohorts that entered between
1969 and 1972 were attracted to the program because they were women

whose family responsibilities meant working limited hours outside the home. As Richter wrote in 1970, "Since the women have homes to keep most are on a part-time study schedule. This sets the pace they will probably be keeping when they have completed their training and are working in Medical Genetics Centers," which usually "are only open one day a week."[54] Of the inaugural cohort (which ultimately included 12 students, as two joined the cohort in January 1970), for example, most of them were mothers. Heimler was 15 years out of college and had four school-age children. The program, whose classes did not start until 10 a.m., suited her as a "mother with young children."[55] Only one member of the first cohort was unmarried and under 26 years of age. At the very beginning of the program, this profile enabled Sarah Lawrence's genetic counseling students and graduates to situate themselves in clinics with little fanfare or professional rivalry. Physicians found it fairly easy to accept these new nurse-like figures that they frequently referred to condescendingly as "girls." These patterns helped to set the stage for the feminization of genetic counseling, a process that has resulted in the field being undervalued in terms of salary and status. The field's heated battles of the 1970s and 1980s, between the more auxiliary title of "genetic associate" (which both Richter and Marks favored) and the more empowering "genetic counselor" (which prevailed), and over what professional body could and should oversee board certification for genetic counselors, reflected the double-edged sword of the feminization of genetic counseling.[56]

Even as these early patterns were leaving discernible imprints on the profession as a whole, the composition of Sarah Lawrence's student body changed. A 1973 site visit of advisors, including Littlefield, Robinson, and Motulsky, praised Sarah Lawrence's program as "the first and best" in the country but underscored that while it "has been successful with older married women, recruitment of young mobile men or women is recommended."[57] Marks moved swiftly in this direction, and under her leadership the cohorts became more diverse in terms of age and region of origin.[58] By the mid-1970s, 62 percent of the incoming class was under 26 and included students from across the country, as well as a small but steady number of men.[59] Whereas Sarah Lawrence's first class was entirely local, almost all from Westchester County, by 1975 nearly half were from outside the New York metropolitan area, a figure that reached 73 percent the following year. The 1976 and 1977 entering classes counted students that hailed from California, Maine, Virginia, Ohio, and Massachusetts.[60] For example, Caroline Lieber, who joined a cohort of 25 students when she started the program at Sarah Lawrence in fall 1978, moved east from the West Coast, having earned her bachelor's degree in science in California.[61] Soon her experience became the norm, not just at Sarah Lawrence but around the country.

More broadly, pathways to master's-level genetic counseling grew over the next five years as additional programs were established at Rutgers University (1971), the University of Pittsburgh (1971), the University of California at Berkeley (1973), the University of California at Irvine (1973), and the University of Colorado Health Sciences Center in Denver (1973). These sister programs extended the field in terms of region, curriculum, and clinical training. Like at Sarah Lawrence, they were often the result of one or two enterprising faculty members who saw the need for an interdisciplinary master's program at her or his institution. As such, these programs sat in different academic homes depending on the particular dynamics of their establishment and each university's administrative structure. Representatives from this initial set of programs gathered at the Asilomar conference center in Pacific Grove, California, in 1974 and 1976 to take stock of the objectives, practices, and issues of genetic counseling. In their second, larger meeting they found that "each program has a different view of the eventual role of its graduates, and that, in fact, we are training a very diverse group of people," adding optimistically, "It is likely that within the range of genetic services required in different settings, each of these roles can be accommodated."[62]

The second program in the country was founded in 1971 at Rutgers University, housed in the Douglass College for women. It was the brainchild of Charlotte Avers, a PhD cell biologist and chairperson of the Department of Biological Sciences.[63] In the early 1970s Avers became interested in establishing a genetic counseling program characterized by rigorous training in medicine and genetics and distinct from the "broader spectrum of emphases" offered at Sarah Lawrence.[64] Searching for a director, Avers consulted several colleagues, who in turn highly recommended Marian Rivas. A recent PhD graduate from Indiana University's Department of Medical Genetics, Rivas had written a dissertation on genetic linkage analyses of thyroxine binding globulin deficiency.[65] Impressed with her study, Victor McKusick offered Rivas a postdoctoral position at the Johns Hopkins University, where she focused on linkage studies of clinical disorders, in particular myotonic dystrophy and tricho-dento-osseous (TDO) syndrome.

Although she was intrigued by Avers's idea, Rivas hesitated at first, primarily because she was not convinced that the complexities of medical genetics could be taught in a two-year-long master's program. After conferring with many colleagues—mainly MD and PhD medical geneticists—to ask for input on curriculum and training, Rivas decided that, with a demanding curriculum, it was possible, and she ultimately accepted the challenge of designing and directing the Rutgers program. In an effort to distinguish their program with on-site scientific and laboratory training, Rivas and Avers put together a team that included cytogeneticist Leonard Sciorra and embryologist Francine Essien.[66] Course work covered the principles of medical genetics, biochemical genetics, clinical genetics, and mathematical genetics, as well as counseling techniques and psychosocial electives. Rivas worried about the prospects of graduates, since "there was no guarantee they would find full-time counseling positions," and thus made sure to prepare her students for positions "in laboratory, clinical or research settings to supplement their counseling activities." In addition, Rivas arranged internships at hospitals in New Jersey, New York City, Philadelphia, Baltimore, and Indianapolis. Opened for admissions in fall 1971, the Rutgers program enrolled 10 students by the following summer.[67]

Unlike Richter, who initially thought that genetic counseling could be a part-time occupation, Rivas felt strongly that the students who trained with her should be employed full-time in some area of clinical genetics where they could practice and promote counseling services. Furthermore, her early cohorts were relatively diverse, including men and women of various ages. Rivas remembers the interviews she conducted with potential students, as she evaluated each applicant's previous course work and general knowledge of biology, life sciences, and mathematics. Rivas viewed course work in psychology as a plus and conducted the interviews "in such a way as to measure his/her reasons for entering the field and to observe whether the applicant could envision and appreciate the varied clinical situations which might arise and the range of emotions and mental anguish that many patients might experience."[68]

Rivas and her students bonded deeply as they put Rutgers's genetic counseling program on the map. Bonnie Baty, a member of the second cohort and currently director of the University of Utah's genetic counseling program, remembers that they had a "sense of being in this together."[69] Rivas has very fond recollections of that period of her life, including the family-style dinners and meaningful conversations she and her students shared: "The Rutgers days were some of the happiest and most reward-ing of my life."[70] Her students held her in the highest esteem, affection-ately calling her the "extraordinary Marian Rivas." Rivas earned praise from colleagues around the country for her innovative program build-ing, and Rutgers's graduates excelled in their own careers as they helped to bolster the visibility of genetic counseling. Despite its deep impact on the early years of modern genetic counseling, the Rutgers program ended in 1980, felled by a lack of university support several years after Rivas had moved to the University of Oregon to collaborate in a multi-institutional study on the mapping of chromosome 1.[71]

On the other side of the country, an inventive program was in the works at the University of California at Berkeley. This new enterprise was housed in the Health and Medical Sciences program, established in 1972 as a stand-alone interdisciplinary unit on the recommendation of a Chancellor's Advisory Committee. Organized to be intellectually flex-ible, this program sought to address "the broader question of health, rather than medicine alone," and fostered the interdisciplinary training of health professionals.[72] Its founders viewed Berkeley's program as an alternative to conventional medical school that engaged multiple health providers and healers and promoted productive collaboration among them: "Our responsibility is to get deeply involved in providing educa-tional opportunities for a diverse group of people who are concerned with health from all points of view."[73] At Berkeley, the Genetic Counsel-ing Option was one of three integrated graduate tracks in Health and Medical Sciences that included a joint medical program with the Uni-versity of California at San Francisco and a mental health program in collaboration with San Francisco's Mt. Zion Hospital. From the outset Berkeley's program described the work of genetic counselors as "social-psychological communication" with "persons or families who have con-cerns arising from genetic diseases or risks of genetics diseases" and with "fellow health professionals and the general public."[74]

Roberta Palmour, now a geneticist at McGill University in Montreal, served as assistant director for Berkeley's program during its early years. She remembers the resounding focus on psychological theories and is-sues, which were central to the program's curriculum from the inception. Rather than psychoanalytic or Freudian in orientation, the program versed students in Rogerian techniques and was fortunate that Erik Er-ikson, a prominent psychologist who had retired from Harvard Univer-sity and lived across the bay in Mill Valley, taught a course in human development.[75] Lucille Poskanzer, who entered Berkeley's program in 1974 after a combination of volunteer and paid positions in women's health organizations, remembers that her instructors accentuated psychosocial and community health, training students in "a different approach to medical care, one rooted in the community that crossed disciplines."[76]

In contrast to Sarah Lawrence, where students rotated among dozens of different facilities in and around the New York area, for Berkeley students the clinical experience was centralized at UCSF's clinics, an arrangement facilitated chiefly by pediatric geneticist Charles Epstein.[77] The initial classes of Berkeley's students also trained with Bryan Hall, a dysmorphologist with expertise in pediatric syndromes and malformations. Open-minded and not beholden to the hierarchical rites of medical training, Hall included genetic counseling students in clinical rotations and patient consultations, treating them "like fellows" alongside the residents and medical students and granting them a great deal of autonomy and trust.[78] In turn, Hall's appreciative trainees sensitized him to the importance of family dynamics and brought effective strategies to working with families in distress into his clinic. Palmour recalls Hall's acceptance of genetic counselors and his realization that they could "work as part of an integrated medical team" that was taking care of patients affected by genetic diseases.[79] For Poskanzer, this meant that she and her peers "considered physicians as true colleagues and partners, so that when we entered professional careers, we challenged the traditional hierarchy of the medical establishment."[80]

In the Rocky Mountain West, Arthur Robinson, a pediatrician and chair of the Department of Biophysics and Genetics at the University of Colorado Health Sciences Center in Denver, arrived in the early 1970s eager to expand clinical genetics. An early supporter of Richter's efforts and a trusted advisor to Sarah Lawrence's program, Robinson approached the Nursing School at the University of Colorado. On the clinical side, he put together a team that included Anne Matthews, a master's-degreed nurse who was hired with the title of nurse geneticist, and Vincent M. Riccardi, a pediatrician with a great interest in cytogenetics.[81] Serving as the University of Colorado's first genetic counselors, Matthews and Riccardi traveled throughout Colorado, Wyoming, Nebraska, and Utah delivering genetic services to families in populated and remote areas.[82] Not hindered by the Nursing School's meager interest in launching a genetic counseling program, Robinson found allies in the Graduate School, which served as the umbrella for the first genetic coun-

Working with a team of children's health and genetic specialists at the Children's Hospital in Oakland in 1977, genetic counselor Lucille Poskanzer, a graduate of the University of California at Berkeley's recently created genetic counseling program, studies a patient's karyotype. In clinics such as these, in the Bay Area and around the country, newly minted genetic counselors found that their unique skill set of psychology and science was highly valued by physicians and patients alike. *Source*: Medical Genetics Unit (Oakland, CA: Children's Hospital Medical Center of Northern California [1977]), p. [10]. Reprinted with permission from Children's Hospital & Research Center of Oakland.

seling program in the Rocky Mountain West. Working with a medical social worker, Matthews played an instrumental role in designing the psychosocial components for the genetic counseling curriculum. The inaugural cohort of three students entered Colorado's program in 1973.[83]

At the University of Pittsburgh, the Graduate School of Public Health provided a home for the founding of a genetic counseling program in 1971. Since its inception, the signature of that program has been a focus on the intersections of public health and genetics, and it was the first program to offer a dual degree, an MS in genetic counseling and MPH in public health genetics.[84] Administrative approval of a genetic counseling program was also granted in 1971 at the University of California at Irvine, which accepted its inaugural cohort two years later. Based in the Program of Social Ecology, this program was the result of an interdisciplinary effort among faculty in medicine, the sciences, and social sciences and initially required course work in social ecology and experimental design. Eventually the program moved to a more disciplinarily appropriate home in the College of Medicine.[85]

Pioneering a Helping Profession

One of the leitmotivs that emerges in oral history interviews with faculty and students involved in the early years of master's degree genetic counseling was their willingness to take a considerable risk on an unknown field that offered little concrete promise in terms of future position or pay. In the words of Heimler, she and her peers were "guinea pigs and pioneers," a sentiment echoed by many of her contemporaries around the country.[86] Michael Begleiter was one of the trailblazers that enrolled in the Rutgers program in 1972, and he believes that his cohorts' participation in Rivas's budding program produced a profound camaraderie that endures to the present.[87] Genetic counselors who matriculated from programs in the early to mid-1970s frequently describe their relationship to the field as a "calling" that enabled them to combine their various interests and talents—in genetics, health care, human psychology, and reproductive issues—into a multidimensional whole. Caroline Lieber, now director of Sarah Lawrence's genetic counseling program, first learned about genetic counseling in a human genetics course she took as an undergraduate at the University of California at San Diego. She was immediately "hooked" and with "no hesitation" knew she wanted to pursue a career in genetic counseling.[88] June Peters, a peer of Begleiter and Baty, pithily sums up a common feeling: "It was just like I was made for this field."[89]

Many genetic counselors explain that their love of science attracted them to the field. In the early 1980s, Barbara Biesecker learned about genetic counseling halfway through her undergraduate career as an unconvinced science/math major at St. Olaf College in Minnesota and "never looked back."[90] Biesecker went on to graduate from the University of Michigan's genetic counseling program and currently directs the joint National Human Genome Research Institute/Johns Hopkins University program. Bonnie LeRoy became fascinated by cell biology as an undergraduate at Albion College, a small liberal arts institution in Michigan. She learned about genetic counseling from an encouraging professor and followed through on his suggestion of applying to the program at

Sarah Lawrence.[91] Since 1989 LeRoy has run the genetic counseling program at the University of Minnesota. Catherine Reiser was riveted when she was first exposed to genetics in high school in rural southeastern Wisconsin in the early 1970s. Her interest in genetic counseling was so great that, at age 16, Reiser sent a letter to a well-known geneticist at the University of Wisconsin. The geneticist offered to meet her at his office, and she traveled to what felt like the big city, Madison, to begin a life-long relationship with the Genetics Department at that university, first as an undergraduate and master's student and later as director of the genetic counseling program.[92] Not only did these women like science, but they were exceptionally good at it, whether it be the statistical aspects or the laboratory work. Nevertheless, most of them were not keen to pursue the demanding track of a PhD or MD, even though some were admitted to prestigious doctoral programs and medical schools along the way. Rather, they wanted, as did Debra Doyle, who attended Sarah Lawrence's program and today is Washington's state genetics coordinator, to be "different than a traditional nurse but less than a doctor."[93]

If genetic science captivated these women, just as paramount was their desire to help people. Carol Norem, who graduated from the University of California at Berkeley's program and today is a leading genetic counselor in the prenatal program at Kaiser Permanente, pursued genetic counseling because she felt that "becoming a doctor or psychologist would require her to focus too exclusively on either medicine or psychology." With genetic counseling she could meld both and, she hoped, provide a service to pregnant patients who all too often only saw doctors who "gave direct commands and medicalized pregnancy."[94] Diane Baker, the first director of the University of Michigan's genetic counseling program, lost her mother when she was a child, an experience that attuned her to grief and trauma. A formative moment on her pathway to becoming a genetic counselor occurred when she was writing a paper for a graduate class in medical genetics she was taking, as a precocious undergraduate, at Michigan State University. While looking through the library stacks for relevant sources, she stumbled across a book about genetic disorders that included photographs of children with related clinical manifestations. As Baker looked at the pictures of these children, she became fixated on the hands that were holding them as they posed. "Whose hands were they? The nurse, mother, physician, assistant?" These unnamed faces occupied her thoughts, and she knew she wanted to meet and help the children and families documented in the book.[95]

Over time, contemporary genetic counseling crystallized into a help-ing profession populated primarily by women, and as such it can be grouped broadly under the rubric of emotional labor.[96] A term coined by Arlie Hochschild in her groundbreaking 1983 study on the taxing emo-tional management performed by flight attendants on the job, emotional labor has evolved beyond its initial applications to service workers and now encompasses professional and expert service work.[97] This concept is particularly useful for understanding the gender and psychological pro-file of contemporary genetic counseling.

Genetic counselors are a unique variant of emotional laborers, car-rying out a kind of work that does not involve hands-on clinical care or medical treatment. Unlike nurses, genetic counselors do not perform bedside tasks that alleviate pain or improve the health status of a patient. Rather, they provide information and a platform for discussion about health conditions for which there frequently is no cure or treatment. Genetic counselors want to do emotional labor, but on a level where they experience more autonomy than workers in other kinds of service posi-tions. Like all emotional laborers, even those of a higher professional sta-tus, genetic counselors easily can become invisible and fall between the cracks, whether in medical institutions or to insurance providers. Fur-thermore, this "emotionally intense work" is psychologically demanding and can take a toll on mind and body. The problem with burnout has prompted some genetic counselors, such as Peters, a Rutgers alumna, to devise strategies for coping, stress management, and locating social support that can be of value to her colleagues.[98]

CHAPTER SIX

Ethics

Shades of Gray in Genetic Counseling

In the late 1960s, as Melissa Richter was drafting her plan for Sarah Lawrence's genetic counseling program, another experiment in institution building was brewing just eight miles away along Westchester County's Saw Mill River Parkway in Hastings-on-Hudson, New York. Like Richter, the philosopher Daniel Callahan, who had earned his PhD at Harvard University, gravitated toward the moral and social issues of reproduction, "family planning and population limitation programs."[1] Also like Richter, Callahan wanted to apply his expertise and intellectual energy beyond the ivory tower, to address the various human dilemmas associated with "the biotechnological developments of that decade."[2] In 1969, the same year that Sarah Lawrence accepted its first cohort of students, Callahan and Willard Gaylin, a clinical psychiatrist at Columbia University College of Physicians and Surgeons, founded the Institute of Society, Ethics and the Life Sciences, soon known as the Hastings Center. This innovative and independent think tank sought to provide an intellectual platform to investigate and discuss *"the social impact of the biological revolution"* (italics in original).[3]

From the outset, the Hastings Center spotlighted genetic counseling and the biotechnological and medical matrix in which it was embedded as an area that demanded sustained attention and analysis.[4] For example, in 1971 the Hastings Center and the John E. Fogarty International Center at the National Institutes of Health cosponsored a conference titled "Ethical Issues in Genetic Counseling and the Use of Genetic Knowledge," which probed the shades of gray associated with the uses and potential uses of genetic information and technologies.[5] Speakers explored a wide range of topics and to a great extent used the question "Where is the future likely to take us in genetics?" as a sounding board to delineate bioethical frameworks for genetic counseling.[6]

One of the center's most eclectic and brilliant minds, Marc Lappé, a toxicologist interested in social and policy questions, struggled with the challenges posed by genetic counseling and medical genetics.[7] The leader of the Social, Ethical, and Legal Issues of Genetic Counseling and Genetic Engineering research group, Lappé was particularly concerned that scientists and specialists involved in genetic engineering and genetic testing were too focused on goals and future outcomes, not on the everyday uses and implications of new technologies such as amniocentesis.[8] Lappé pushed genetic health providers to be aware of the values embedded in the disease prevention model of genetic testing and cautioned against implicitly or explicitly privileging the gene pool or society over the individual. In one pointed essay published in the second issue of the *Hastings Center Report*, launched in 1973, Lappé asked "the genetic counselor: responsible to whom?" and responded, "I think that genetic counselors may be misguided if they feel that their ethical obligation is in *any way* to future generations" (italics in original).[9] He questioned applying the utilitarian impulse pervasive in Western philosophy—to do the greatest good for the greatest number—to genetic counseling and instead proposed the Eastern philosophical concept of *ahisma*, continuous obligation to minimize potential suffering, as a guiding principle. Although one could caricature his distinction between Western and Eastern philosophy, Lappé was farsighted in his suggestion that genetic counselors should embrace a situational ethics undergirded by human caring and empathy.

Today genetic counselors wrestle with many of the questions raised by Lappé 40 years ago, particularly when they strive to put the fundamental ethical principle of autonomy into practice with their clients. Respect for patient autonomy has become something of a mantra in genetic counseling, and in medical treatment more broadly. While it is a noble goal, autonomy is neither transparent nor static and can be difficult if not impossible to operationalize equally in all genetic counseling scenarios.[10] Indeed, the power and limits of autonomy are tested in the real-life situations of genetic counselees, who make decisions not in a neutral vacuum but within the constraints of their social, economic, cultural, and personal circumstances. Advances in genomic medicine, changing patient demographics, and shifting cultural, religious, and political values also constantly reframe the reach and relevance of autonomy. Furthermore, some clients cannot access or afford autonomy.[11] In the twenty-first century the great challenge is how to activate the ethos of self-determination and patient respect associated with autonomy, considering the wide range of diagnoses, results, and personal and family situations.[12]

Little recognized in the scholarly literature is the extent to which the development of the concept of autonomy was intertwined with medical genetics and genetic counseling. Yet, a key strand of the historical genealogy of autonomy leads us back to the genesis of medical genetics and genetic counseling. Moreover, studying the rise and consolidation of autonomy through the prism of human genetics demonstrates the vital roles played by psychotherapy, gender dynamics, and the gestation of bioethics in this process. Understanding autonomy in genetic counseling requires an exploration of its turbulent relationship with nondirectiveness, a principle that has been central to contemporary genetic counseling.[13] Since the 1990s, there has been heated debate among genetic health care providers about the value of nondirectiveness, both as a general method and in light of the communicative expectations of genetic counseling's diverse clientele.[14] For many genetic counselors, dissatisfaction with nondirectiveness has intensified into calls to abandon it altogether.[15] At the same time, genetic counselors' commitment to autonomy remains strong and resilient.[16]

How did nondirectiveness and autonomy become interwoven into genetic counseling? This process can be traced back to four overlapping and temporally staggered paths. The first is Carl Rogers's development of nondirective and client-centered counseling in the early 1940s, which he formulated as an alternative to the more instructive and didactic models associated with Freudian psychoanalysis. The second is Sheldon Reed's partial and incipient advocacy of autonomy and his choice of the label "client," nomenclature he almost surely borrowed from Rogers. The third is the cohesion of autonomy as a tenet of American bioethics in the late 1960s and early 1970s, a period in which physicians, theologians, community activists, and other interlocutors challenged the status quo in biomedicine and created and codified new principles and protocols for patient care and medical research.[17] Finally, beginning in the 1970s and still hinged to nondirectiveness, autonomy became ensconced in the conceptual warp and woof of genetic counseling training programs and professional organizations. Genetic counselors, preponderantly women who strongly endorsed reproductive rights, incorporated bioethical principles into their position papers, mission statements, and practice guidelines even as some of them recognized the need for conceptual flexibility in their daily counseling practice.

Carl Rogers and the Transition from Nondirective to Client-Centered Counseling

In the late 1930s Carl Rogers, a clinical psychologist trained at Columbia University's Teacher's College, was working at the Child Study Department of the Society for the Prevention of Cruelty to Children in Rochester, New York.[18] Seeing a steady stream of children and adolescents with serious psychological problems, many of whom suffered from mental illness and had been brutalized in violent home situations, Rogers was searching for therapeutic approaches that were thorough, scientific, and sensitive to the mosaic of family, environment, and personality. Rogers admired Sigmund Freud's insights into the human unconscious and the complexity of psyche and emotion but was unimpressed with Freudian psychoanalysis, which he found neither useful nor practical. It took too long, demanded interpreting the analysand's life through an "elaborate theoretical superstructure," and usually involved a directive if not imperious analyst.[19]

Reflecting his combination of pragmatism, belief in the centrality of an individual's immediate life circumstances, and reverence for the uniqueness of the self, Rogers gravitated toward Jessie Taft's "relationship therapy" and, more significantly, Otto Rank's "will therapy." Rank's approach assumed that individuals, no matter how psychologically compromised, possessed the capacity of self-direction and the creative powers, or "will," for transformation. More so than Rogers, Rank elaborated his theories in opposition to Freud, opining that the primal trauma experienced by humans was rooted not in the psychosocial turmoil of early development but in the cataclysmic moment of birth.[20] For Rogers, Rank's approach offered greater flexibility than the Freudian model and sought to engender a dynamic and interactive psychotherapeutic process in which the individual came to possess her or his own skills of self-reflection and improvement. As Rogers explained in an interview years later about this formative period in his life, "I came to believe in the individual's capacity. I value the dignity and rights of the individual sufficiently that I do not want to impose my way on him."[21]

From this stepping stone, Rogers developed the theory of nondirective counseling, which he believed was structured by three major phases: (1) release, (2) insight, and (3) positive action based on insight. To implement objectivity and scientific method, Rogers devised the technique of audio-recording psychotherapy sessions, which he then transcribed and annotated following a rigorous and systematic formula.[22] Indeed, one of the first case books in American psychology to contain transcribed and analyzed counseling sessions was dedicated to nondirective counseling. This volume sought to show that this vein of counseling was noncoercive and nonargumentative and sustained a therapeutic platform for counselees to become aware of their own problems and locate the animus for change within. It included five distinct cases presented by different psychologists. Rogers's case involved a young woman near suicide whom he successfully and gradually counseled toward self-empowerment and self-acceptance.[23] To work with her, Rogers employed verbal techniques of clarification and repetition rather than reframing and advising, and he spent much of the session listening, not talking.

Calling his approach "nondirective" helped Rogers to disassociate himself from the strictures of Freudian psychoanalysis and underline the paramount role of counselee. However, Rogers still found the term lacking, replacing it several years later with "client-centered," which he

Carl Rogers in a counseling session with a client at the University of Chicago, circa 1950. The seating arrangement, in which Rogers and his client sit comfortably across from each other at a table, differs from classic Freudian couch therapy and captures the interactional style of client-centered counseling. *Source*: Natalie Rogers. By permission.

preferred for several reasons. First, as Rogers explained in an interview, he had "never been satisfied with the term *patient*," given that he was a clinical psychologist with a PhD, not a psychiatrist with an MD.[24] In addition, to be a patient was to have a pathological condition, a description that was often inaccurate and stigmatizing for counselees. Second, "nondirective" referred to the type of therapy, not the living, breathing person engaged in the therapeutic process. The term "client-centered" honored the counselee and confirmed that he or she stood at the center of the psychotherapeutic exchange. As Rogers explained, "A client is self-responsible, going to someone else for help but still containing the locus of evaluation within himself, the locus of decision. And he's not putting himself in the hands of someone else. He still retains his judgment. It seemed like the best term I could find."[25]

Over the course of the 1940s and 1950s, as he moved from Rochester to Ohio State University and then to the University of Chicago, Rogers expanded and elaborated his client-centered therapy, which soon became one of the pillars of American clinical psychology.[26] At the crux of Rogers's approach was a belief in the autonomy of the individual, who possessed the tools of self-examination and reflection that could lead to insight and transformation. In the 1960s, Rogers relabeled his approach "person-centered" and established the Center for the Studies of the Person in La Jolla, California.[27] For the last two decades of his life, until his death in 1987 at the age of 85, Rogers traveled the world to promote person-centered workshops and encounter group therapy. He stood at the forefront of humanistic psychology, advocating a brand of therapeutic self-determination that resonated with the countercultural and progressive political currents of the era.[28]

Although Rogers did not use the word "autonomy" on a regular basis, his psychotherapeutic approach was built on what Ruth Faden, Tom Beauchamp, and Nancy King have identified as the three key conditions of autonomous decision making: (1) intentionality (actions are willed and performed according to one's plan), (2) understanding (actions based on understanding of the situation and choices), and (3) voluntariness (actions are made without controlling influences).[29] From his early career working with troubled children to his final years spearheading humanistic psychology, Rogers always placed a premium on self-respect, dignity, and expression, traits that a well-trained and attuned counselor could encourage but not compel from her or his client.

Sheldon Reed's Equivocal Approaches to Genetic Counseling

If Rogers developed a new form of psychotherapy that was explicitly nondirective and client centered, then Sheldon Reed applied Rogerian terminology to an emergent nexus of medicine and psychology where the counselor's intentions were much murkier. In 1947, Reed coined "genetic counseling" and began offering services at the University of Minnesota's Dight Institute. In 1974, Reed asserted that he founded genetic counseling as a "kind of genetic social work without eugenic connotations."[30] However, Reed's assessment of his essential role in ensuring that genetic counseling was ethically pristine was only half right.

Similar to his relationship to race and disability (explored in chaps. 3 and 4), Reed's approach to genetic counseling was complicated and seemingly inconsistent. As Diane Paul and Molly Ladd-Taylor both have suggested, Reed believed that most of the middle-class white clients who sought out genetic counseling possessed the capacity for rational decision making and could act as autonomous agents.[31] He also recognized that the ultimate outcome of genetic counseling might be more dysgenic than eugenic insofar as it prompted couples to have more children with expressed or unexpressed genetic conditions. Often these included couples with autosomal recessive conditions traceable in their family trees that chose to proceed with pregnancy, likely judging the 1 in 4 chance of a homozygous recessive child as worth the "risk."

Reed's ambivalence was reflected in the tripartite requirements he listed for physicians or PhD geneticists to provide genetic counseling. The first, basic genetic knowledge, was technical, informational, and indispensable to the provision of as accurate a diagnosis or risk assessment as possible. The third, "the desire to teach," encapsulated the benevolent paternalism of his generation of medical experts. It was Reed's second requirement, "a deep respect for the sensitivities, attitudes, and reactions of the client," that resonated strongly with the Rogerian ethos and represented an early advocacy of autonomy in genomic medicine.[32]

More so than most other medical geneticists practicing genetic counseling, Reed instructed providers not to compel or coerce their clients. As he asserted in a 1952 panel discussion on genetic counseling sponsored by the American Society of Human Genetics, "It is an extremely rare case, which, in the experience of the Dight Institute, can be advised outright, as to whether or not reproduction is indicated. The counselor has never suffered the particular circumstances which the parents of the affected child suffered and therefore cannot completely understand their feelings."[33] Twelve years later, at a symposium on human genetics in public health, Reed explained that "the geneticist cannot indulge in directives, as there is always a possibility that the next child of any couple may be abnormal."[34] During the same time period, Reed stated, "We do not try to sell anybody anything, that is, we are not telling them they should have zero or twelve children. We are not trying to persuade them to do anything."[35] It was with respect to the communication of genetic risks that Reed most consistently asserted the "cardinal rule of counseling," namely, that the "expert merely lays all the information before the prospective parents and explains the risks, if any. He doesn't recommend whether they should have a child or not." Since the genetic counselor had not "experienced the emotional impact of their problem, nor is he intimately acquainted with their environment," he was not in position to make a categorical decision on behalf of any client.[36]

Reed stood in contrast to his predecessor at the Dight Institute, Clarence P. Oliver, who moved to the University of Texas at Austin in 1946. Oliver was a brazen advocate of the eugenic role of genetic counseling and emphatically stated that counselors should give forceful advice in cases where genetic defects were likely in future offspring. For example, Oliver expressed no qualms about exaggerating or withholding information from a father with suspected hereditary diabetes who had one daughter with diabetes and was considering a second child. Based on the working hypothesis that any potential offspring had a 1 in 9 chance of inheriting diabetes (little was known in the early 1950s about genetic predisposition to type 1 diabetes), Oliver stated that further procreation would constitute a gamble "taken at the risk of child." Thus, he believed it was necessary to "do more than just give the probability that the good or bad will occur" by making the "picture as dark as possible to help" the father decide not to attempt any further procreation.[37]

From the 1940s to the 1960s, Oliver's approach was not uncommon. F. Clarke Fraser, a geneticist at McGill University who helped to develop genetic counseling, remembers that when he started "the traditional counseling was very directive." For example, his contemporary Madge Macklin would not hesitate to tell a couple she judged to be at an elevated risk of a hereditary disorder "a resounding no!" during a consultation. For Fraser, this model was too extreme, but he was also unnerved by the "reaction that you mustn't be directive at all," an injunction that sometimes conflicted with his physician's mind-set and the patients' desire for medical advice.[38]

Tellingly, Reed could be much more directive when he judged someone to be incapable of rational decision making. For Reed, such individuals included patients in homes for people with mental retardation or members of families with histories of Huntington's disease. Reed believed that their diminished cognitive capacities and inferior genetic worth justified if not required public and expert intervention. Ultimately Reed divided the world into those of lesser and greater genetic value, a distinction that generally aligned with capacity for rational cogitation: "There is no question in my mind but that the 25,000 students at the University of Minnesota are of greater genetic value to Society than the 25,000 patients in our Institutions for the mentally incompetent."[39] In 1951, Reed responded to a St. Paul resident about the future of Minnesota's sterilization law, which had been passed in 1925 largely as a result of pressure from Charles Fremont Dight, the benefactor of the Dight Institute. The local resident was concerned about ongoing attempts to repeal the law, an outcome that Reed and the Minnesota Human Genetics League strongly opposed.[40] As late as 1968, Reed supported prophylactic sterilization. Writing to a colleague that year in support of a revision to Oregon's sterilization law, he endorsed the long-standing eugenic belief that sterilization should be a condition for release from a feebleminded or psychiatric home. Oregon's proposed legislation more explicitly provided a mechanism for "those who used to be in institutions but are now circulating in the general public, or are no longer under continuous supervision." Reed touted the revamped law for offering "protection for this group."[41] Thus, Reed and many of his colleagues reserved autonomous decision making for well-to-do clients who would act in the benefit of future generations based on a sense of societal duty.[42] Reed described his clients as mainly "stable couples in closely bound family units."[43] Unlike these ideal and intelligent young couples, "the common garden-variety of moron family without cultural motivation" was not "likely to come to anyone for counseling about the rampant mental retardation in the family" because they simply didn't care.[44]

Nevertheless, within the parameters of his two-tiered approach and lingering eugenic ideology, Reed provided a substrate for individuality and autonomy in genetic counseling.[45] One of the most important ways he did so was by choosing the term "client" over "patient," an interesting and quite conscious decision given that most of the counselees came to the Dight Institute because of medical conditions and family health concerns. As Reed told the Sacramento Genetics Institute in 1967 in discussing the professional profile of genetic counselors and their work, "The word client is used, rather than patient, to describe the person who comes for genetic counseling because the process involved is much more one of helping than of healing."[46]

Like Rogers, Reed held not an MD, but a PhD. A careful review of the Dight Institute for Human Genetics Records and Carl Rogers Papers, housed at the Library of Congress in Washington, D.C., does not reveal any written or verbal communication between the two men.[47] Thus, it remains something of a mystery why or how Reed elected to use "client," but it was most likely through general familiarity with Rogers's popular and widely circulated texts. Whatever the route to the term "client," Reed's decision had significant implications for the development of the field of genetic counseling. Until the 1960s, however, autonomy and nondirectiveness as formulated by Reed were stymied by the eugenics paradigm at places like the Dight Institute and Wake Forest University. It would take the social transformations attached to the civil rights era, particularly as a new generation of primarily female genetic counselors began to remake and populate the field and with the explosion of genetic and reproductive technologies, for these concepts to acquire their current bioethical meanings.[48]

Autonomy at the Core of American Bioethics

During the early 1970s, patient autonomy, reproductive choice, and voluntary decision making became hallmarks of bioethics and had particularly strong resonances in genetic counseling and medical genetics. Historians have recognized the role that medical advances such as organ transplantation and the availability of long-term life support systems played in spawning uncharted areas of moral deliberation. These new technologies appeared in hospitals as trust in doctors was diminishing and as revelations about human experimentation on vulnerable subjects in prisons and state schools were coming to light. The convergence of these factors fostered the creation of new areas of moral and medical inquiry, bioethics, and a new professional—the bioethicist—was born, charged with providing guidance about medical research and interventions.[49] Bioethics arose in the context of the civil and human rights issues being debated and formulated in political, legal, and community domains. Keen to apply the concepts contained in the Nuremberg Code and other international documents concerned with human experimentation and patient protections, bioethicists addressed medical abuses and attendant conundrums in American society.[50]

One of the hotbeds of such conversations was the Hastings Center, which stood at the forefront of thoughtful and rigorous reflection on the implications of new genetic and reproductive technologies and the amplifying role of genetic counselors. As part of its first year of interdisciplinary talks, the Social, Ethical, and Legal Issues of Genetic Counseling and Genetic Engineering research group, led by Lappé, focused on four issues that could facilitate discussions related to "the ability to anticipate genetic disease before birth."[51] This quartet included eugenics versus therapy, effects on the parents, privacy, and coercion. Overall, this group probed the dilemmas that came with the emergence of genetic tests and more genetic information for mothers and families making reproductive decisions and for members of society concerned about the potential for "playing god."[52] Lappé was particularly interested in the implicit values of "normal" and "abnormal" that undergirded some of the impetus for prenatal testing. Waving a cautionary flag about the potential problems associated with amniocentesis, which was just becoming more widely available, and its moral and social ramifications, he noted in 1973 that "there will inevitably be those who, armed with cost-benefit analysis, demonstrate that it is always easier and cheaper to terminate a 'genetic illness' than to treat the child burdened with it—and cheaper still to prevent the conception of a 'defective' in the first place than to have to live with the burden of the person who embodies the defect."[53]

In addition to providing a forum for the intellectual inquiries posed by Lappé and colleagues, the Hastings Center set a high standard by "insistence, even then, on getting the facts right."[54] During the 1970s, many observers and critics worried aloud about the prospects of genetic engineering and the coming of an alternately liberating or shackling brave new world. To the Hastings Center's credit, its scholars paid careful attention to the fine and technical details of newly launched genetic screening programs and new discoveries such as polymerase chain reaction (PCR) and techniques of recombinant DNA. Over the ensuing years, as discoveries and developments in medical genetics occurred at a brisk and sometimes vertiginous pace, the center provided a forum for debating the strengths and limits of autonomy in genetic counseling, particularly when it came to issues of disability, difference, and the power dynamics of decision making.[55]

Discussions about the moral parameters of new genetic and reproductive technologies occurred in other venues as well, most notably the Joseph and Rose Kennedy Center for the Study of Human Reproduction and Bioethics, founded in 1971. One of its first sponsored activities was the international symposium "Choices on Our Conscience," held at the John F. Kennedy Center for the Performing Arts in Washington, D.C. On the morning of October 16, more than 1,200 attendees crowded into the morning plenary session, which began with a screening of the 30-minute film *Who Should Survive? Is Survival a Right?* This film, which featured actors, documented two cases at the Johns Hopkins University, in 1963 and 1971, when children born prematurely with trisomy 21 did not receive surgery for duodenal atresia, an intestinal obstruction that could be treated successfully with surgery.[56] In the first case, the mother, who already had two children, conveyed to the pediatrician that "she would rather have the baby die and end it all than to undertake care taking of a severely handicapped baby."[57] The attending physician reviewed this difficult case, explaining that it was apparent that the parents "had already made up their minds and did not wish the child to be treated. I personally feel that this decision of theirs is unwise in practice, even though right in principle; inevitably they must bear the guilt of having produced a defective child. The added burden of becoming the agent of that child's death is, I feel, more than they will be able successfully to bear and to overcome."[58] In the second case, the parents also declined surgery and their newborn boy expired 13 days after birth.[59]

Robert Cooke, a renowned pediatrician and father of two disabled children (see chapter 4), was instrumental to the making of *Who Should Survive?* and hoped that its screening would foster discussions about complex ethical issues involved in these kinds of cases.[60] As he wrote to one local mother who had written to him, distraught after having seen the film, "I made the film on the mongoloid baby with the help of the Joseph P. Kennedy, Jr. Foundation because in many places life and death decisions are based on a mistaken belief in the quality of life. I believe that for a society to survive it must bear responsibility for its least capable, not its most competent."[61] *Who Should Survive?* garnered a great deal of attention in the media. Walter Cronkite remarked in a news commentary, "One would think in a civilized society there would be a less barbarous approach to taking care of the helpless."[62] After it was broadcasted, Cooke received countless letters from across the country, mainly from mothers, outraged by what they had seen. The mother of a child with Down syndrome in Parsippany, New Jersey, wrote that she, like other parents, wanted more than anything to produce "perfect" children. Her initial disappointment of having a child with Down syndrome prompted her to institutionalize her daughter, but after three months she and her husband decided to rear her at home. She felt that she had made the right decision, going on to say that it isn't "always easy or fun but it is rarely hard or sad anymore and there is much real joy and satisfaction in her apparent happiness."[63] Another mother wrote, "I simply cannot comprehend that such atrocities are taking place today and in such respected institutions as Johns Hopkins Hospital. . . . When I read the atrocities committed by the Nazis against children, I was stunned. When I read of parents beating their children to death, I am shocked. When I read of Calley's troops at My Lai murdering innocent children, I was deeply troubled. But when I read of what happened to an infant boy at Johns Hopkins Hospital, I am horrified!"[64]

Who Should Survive? did not resolve the thorny ethical issues at hand, but that of course was part of Cooke's motivation for making the film—that it would encourage wide-ranging discussion and reflection about values and decisions related to disability, the potential limits of parental rights, and the appropriate role of physicians in saving or prolonging life.[65] This provocative program, as well as the outcry it produced, played a role in shifting attitudes toward people with Down syndrome and fostered a host of new developments, such as the formation of bioethical review committees at many hospitals across the country to deliberate on vexing cases related to reproduction, birth, and genetic disorders.

During the same period too, the Johns Hopkins Hospital served as the site for research projects on the ethical issues of the new technologies of prenatal diagnosis. For example, in 1977 John Fletcher, who was affiliated with the Hastings Center, contacted the Prenatal Diagnostic Center, founded in 1969 as one of the first clinics in the country to offer amniocentesis. Fletcher was "particularly concerned with issues like fetal research, prenatal diagnosis, impact of abortion on patient and society, etc." In his capacity as head of a new program at the National Institutes of Health, Fletcher wanted to study these issues at the Johns Hopkins Hospital because of its high patient load and excellent record keeping. Richard Heller, then director of the Prenatal Diagnostic Center, responded to this request not with defensiveness but with interest, telling him, "As you may know, I am intensely interested in the philosophical and moral issues surrounding prenatal diagnosis in particular and medicine in general, and this is a study which I have been longing to do for years."[66] Fletcher's research eventually resulted in the formulation of a set of guidelines for the ethical, social, and legal issues in prenatal diagnosis, which stressed patient autonomy and confidentiality, clear division between clinical and research activities, and counseling that was "non-coercive and respectful of parental views about abortion."[67]

Genetic Counselors and Bioethics

Master's-level genetic counselors were settling into their new jobs at clinics around the country in the early to mid-1970s as hearings in the U.S. Congress addressed the violations of human rights and bodily integrity, not to mention racism and discrimination against people with disabilities, that occurred during the 40-year-long Tuskegee Syphilis Study and with the involuntary sterilizations of thousands of minority and poor women in the late 1960s and early 1970s. Cognizance of these abuses, as well as the long and dark shadow of Nazi genocide, shaped ideas about the goals and limits of genetic counseling. For instance, Arno Motulsky of the University of Washington, an advisor to Sarah Lawrence's program, published a penetrating article that same year that evaluated the ethical responsibilities of physicians with regard to reproductive and genetic medicine. Looking back at an often ugly past and forward to an uncharted future, he wrote that "enforced sterilization should be strongly rejected" and "open discussion and freedom from coercion are the best guarantees for ultimate success."[68] Motulsky encapsulated the ethos of genetic counselors, whose duty was to "put the interests of the patient and his family before those of society and the state. The genetic counselor pursues medical and not eugenic objectives."[69]

The first systematic study of genetic counseling, conducted by James R. Sorenson and colleagues at Boston University School of Medicine in the late 1970s, demonstrated that genetic counselors rejected directiveness. Analysis of more than 2,000 genetic counseling sessions at 47 clinics, as well as of questionnaires produced for genetic counselors included in the study, found that "almost all counselors do not feel that their role includes directly advising or telling clients what to do."[70] For example, 98.6 percent of their interviewees agreed with the statement that their role was to "suggest that while you will not make decisions for patients you will support any they make," and 93.1 percent with the statement "tell patients that decisions, especially reproductive ones, are theirs alone and refuse to make any for them."[71]

The founder of the country's first genetic counseling master's program, Melissa Richter, arrived at genetic counseling partly because of anxieties about overpopulation. Nevertheless, she was a steely supporter of individual reproductive choice. Revealing in this regard is Richter's correspondence with Jérôme Lejeune, who identified trisomy 21 as the cause of Down syndrome in 1959. Lejeune was a devout Catholic who opposed abortion. In his 1969 William Allan Memorial Award Lecture to the ASHG, Lejeune asserted that scientists should be guided first and foremost by the sentiments of "humility and compassion" when responding to the lives and needs of people with disabilities, stressing that "even the most disinherited belongs to our kin."[72] In response to the existential question (What is human?) Lejeune considered in his lecture, Richter posed the counterquestion (What is not human?) in a letter to him. She told Lejeune that she believed that "parents have the right not to run the risk of having two, or three or five children with cystic fibrosis, but that society at this point should not legislate that no one should have a child with cystic fibrosis." In Richter's perspective, which was aligned with the feminist health movement, reproductive decisions were private and should not be dictated by the state.[73] More broadly, Richter pondered in letters and lectures the assumptions intrinsic to the disease prevention model, how to communicate genetic risk with sensitivity and scientific accuracy, and what function public agencies should play in genetic testing and genetic counseling. In January 1972, she responded to students who expressed "a desire to discuss the responsibility of exerting influences on the lives of others" by organizing a workshop on the ethical, moral, and social implications of genetic counseling.[74]

To this day, the influence of the formation of master's-level genetic counseling programs on the emergence and priorities of bioethics remains a largely overlooked topic. Yet grappling with the bioethical dilemmas raised by genetic testing was something genetic counseling students and practitioners did (and continue to do) every day. At the outset of Sarah Lawrence's program many students wanted forums to discuss the moral conundrums they confronted, especially with regard to conveying information to parents about probable or identified defects in their child. Yet when Richter designed the curriculum for Sarah Lawrence's program, she included only one course related to psychology and counseling. To complement the students' science and medical course work, Richter added a class on social psychiatry. However, instead of presenting theoretical or applied psychotherapy, this course interrogated the general human condition; required authors were John Donne, Bertold Brecht, and George Orwell, not those affiliated with any method of clinical psychology.[75]

When Joan Marks became codirector of Sarah Lawrence's program in 1972, she was passionate about integrating the psychosocial component into the curriculum, adding Columbia University psychiatrist Stephen Firestein's seminar in psychology. This course, which was offered from 1972 until the late 1970s, taught students to grasp an "expanded understanding of who the client is" by learning about family and ethnic background.[76] Firestein's syllabus included texts on interviewing styles in social work and family therapy; works in the psychoanalytic canon by Sigmund Freud, Annie Reich, and Karen Horney; and up-to-date articles on ethical issues related to abortion, mental retardation, and sterilization. His approach can be described as psychoanalytic, with a strong but not dogmatic Freudian influence, and designed to train students in the psychodynamic dimensions of risk, anxiety, grief, loss, and stigma.[77] Perhaps most important, Firestein's course meant that for the first time genetic counseling students recorded and critically analyzed mock interviews in order to hone their skills. Elsa Reich, who entered Sarah Lawrence's program in 1972, remembers Firestein's effective teaching style, as he helped students cultivate an empathic approach to interacting with clients with as few preconceptions about them and their circumstances as possible.[78]

In 1978, an explicitly Rogerian course, "Client-Centered Theory of Personality and Therapy," offered by Sarah Lawrence psychology professor Marvin Frankel was added to the curriculum.[79] Taught until the early 2000s, this course relied almost entirely on works by Rogers and provided extensive training in "tape-recorded client-centered interviews with role-playing clients."[80] Frankel contrasted client-centered and psychoanalytic approaches, emphasizing Rogers's verbal techniques of clarification, attention to the client's unique phenomenological framework, and unwavering respect for the client's right to create his or her own narrative trajectory.[81] Frankel sought to teach students deep empathic listening, "which was in stark opposition to the mere repetition and rephrasing which was no more than a misguided caricature" of an authentic counselor-counselee interaction.[82] Even though Frankel acknowledged that the use of genuine Rogerian techniques might not lead to a concrete resolution during a brief counseling session, he believed that genetic counselors should try to seed a "self-reflective attitude that could enable the client to resolve the issue at hand after the interview." Frankel sought to teach students that the genetic counselor's job was to listen "empathically with total acceptance and the absence of a skeptical or analytic point of view" and to facilitate a psychological interaction in which the clients were "given the opportunity to collect their thoughts and reduce their arousal level to a more optimal state and so make their own decisions based entirely on their own thought."[83]

As did Firestein and Frankel for their courses, Joan Marks incorporated ethical issues into her three-semester course "Issues in Clinical Genetics." Marks's capstone course covered a wide range of readings on abortion, disability, race, ethnicity, and reproductive rights.[84] Many of these readings were authored by bioethicists affiliated with the Hastings Center, including Lappé, who lectured to genetic counseling students on

Genetic counseling students learn counseling techniques and discuss ethical and psychosocial implications of communicating genetic information, Sarah Lawrence College, most likely in Joan Mark's capstone course "Issues in Clinical Genetics," circa 1985. *Source*: Sarah Lawrence College Archives.

ethical issues in 1972 and returned in spring 1975 to teach a course in developmental biology.[85]

The "client-centered" model sat at the core of most emergent genetic counseling programs, where directors such as Marks and Seymour Kessler at the University of California at Berkeley wanted to bolster the psychosocial training of their graduate students. For Kessler, this meant disassociating genetic counseling completely from earlier eugenic and directive models that privileged the population over the individual.[86] However, it also meant abandoning the notion that nondirectiveness was useful for a meaningful genetic counseling session. Kessler put this perspective into practice as the director of Berkeley's program, whose founding tenets included collective health and holistic paradigms of healing. Kessler employed Rogerian techniques of audio recording and role-playing, as he strove to instill in students a deeper understanding of both the counselor and the counselee as human beings working through what likely were painful and difficult feelings related to genetic disease and disability.[87] Starting in the 1990s, Kessler led the charge against nondirectiveness, a principle he called "crap" and viewed as a superficial understanding of human psychology not suited to people with complex human needs. Instead of worrying about not engaging with clients, genetic counselors should simply ask, "How can I help this human being?"[88] Kessler agreed that when practiced effectively nondirective counseling had the potential to promote active, engaged counseling in which the client arrived at the appropriate decision by him- or herself.[89] For the most part, however, nondirectiveness was incomplete or misguided.[90] Seeking to transcend the impasse between directiveness and nondirectiveness, Kessler combined psychodynamic and rational-cognitive models with the aim of providing scientifically accurate and emphatic genetic counseling to clients.[91]

A leader in the field who wanted to disseminate his critiques, Kessler challenged nondirectiveness for being vapid and duplicitous in a series of influential articles that appeared in the *Journal of Genetic Counseling* and the *American Journal of Medical Genetics*. In one article, Kessler analyzed the transcript of a counseling session involving a pregnant woman and her husband. Because she was 40 years old, falling in the category of advanced maternal age, the wife was referred to a genetic counselor to determine whether she and her husband wanted amniocentesis. On the surface, the counselor's approach was nondirective insofar as he offered facts and information. Kessler and a trusted colleague scrutinized the session transcript and concurred that the counselor failed to be truly client centered, privileging content over process. They also criticized the counselor for presenting, with medical information and images, a decisively negative picture of trisomy 21, or Down syndrome, the chief chromosomal anomaly detected by amniocentesis. Kessler's exquisite analysis of this session demonstrated that the counselor had been "most directive in that his interventions have been aimed (consciously or otherwise) at influencing the counselees' behavior so that they will agree to have the amniocentesis procedure."[92]

Publications and student training at the Berkeley program foreshadowed the rising dissatisfaction with nondirectiveness that crested in the 2000s as a slew of studies and review articles contended that this foundational precept was neither possible nor desirable. According to critics, nondirectiveness permitted genetic counselors to get off the ethical hook or impeded them from truly engaging with clients.[93] Jon Weil, who became the director of Berkeley's program in 1989, several years after Kessler resigned, never stressed nondirectiveness, instead promoting active counseling skills and "attentiveness to the medical-genetic issues confronting the client and his or her personal, psychological and social needs, wishes and values."[94] For Weil and many other genetic counselors, nondirectiveness was valuable insofar as it clarified that genetic counseling should be noncoercive and nonprescriptive and reject the eugenic model of advice. More critically, Berkeley's program from its inception included a course "on ethical dilemmas, dilemmas of the type that are intrinsic to most genetic counseling situations in which human rights, societal demands and individual emotional needs often come into conflict."[95]

Bonnie LeRoy, director of the genetic counseling program at the University of Minnesota, encourages future graduates to appreciate that they have a particular kind of medical expertise that often cannot be communicated effectively or sensitively to patients utilizing primarily a nondirective model.[96] After earning a master's in genetic counseling at Sarah Lawrence College, LeRoy arrived to Minneapolis in the 1980s as Reed's long directorship was coming to a close and the Dight Institute downsizing until, in 1991, it ceased to have any attached faculty.[97] LeRoy's interest in establishing a master's degree training program was bolstered by an emphasis on bioethics and genomics among her academic colleagues who were aware of the Human Genome Project and its implications. Building on this momentum while recognizing the constructive imprints left by Reed and the Dight Institute, LeRoy took genetic counseling at the University of Minnesota into an era concerned with emergent concepts of risk assessment, patient privacy, and confidentiality, as well as the ramifications of "having knowledge of some of our genetic *predispositions* to health problems" (italics in original).[98]

In the end, many genetic counselors have come to the conclusion that nondirectiveness can easily constrain them by emphasizing what they should *not* do, thus limiting "the role of the genetic counselor to that of information provider" and inhibiting "the use of the full range of relevant counseling techniques."[99] In *Early Warning: Cases and Ethical Guidance for Presymptomatic Testing in Genetic Diseases*, an insightful book with 29 cases of genetic counseling sessions, the majority related to HD, a leading group of genetic counselors reject nondirectiveness based on their combined decades of counseling experience: "Neutral, flatly presented 'nondirective' information places the consultand's decision in a vacuum and ignores what the counselor knows about the suffering associated with HD, the social reaction to abortion, and its possible psychological effects."[100] Marks states that while nondirectiveness played a critical role during the inception of genetic counseling training programs, "over time it evolved to become a troublesome word." She believes that patient autonomy is much more important.[101] Barbara Biesecker, director of the genetic counseling program of the National Human Genome Research Institute / Johns Hopkins University, asserts forcefully that nondirectiveness is "not helpful" and has contributed to a "distancing effect on the counselor-client relationship" and a lack of engagement with emotional issues.[102]

In tandem with the turn away from nondirectiveness, genetic counselors have underscored the importance of autonomy, which actually was the principle at the forefront of discussion and debate about actions and rights of clients during the formative years of contemporary genetic counseling in the 1970s. When Luba Djurdjinovic entered Sarah Lawrence's program in 1976, she realized that her motivations were largely to "right a wrong." Born to a Serbian father and a stateless mother in a displaced persons camp in Germany after World War II, Djurdjinovic moved with her family to Harlem in the 1950s. She grew up with very limited resources, but with parents who placed a high premium on education. After a very successful undergraduate and postcollege career working in biology and histology laboratories, one day in the local library

Djurdjinovic read about genetic counseling for birth defects and was very intrigued. In due time, she found the program at Sarah Lawrence, was accepted after an on-campus interview, and embarked on her career.

For Djurdjinovic, it was important to recognize that the roots of her newly chosen profession could be traced to the eugenics movement. She was painfully aware of the horrors of World War II and, although not Jewish, had experienced firsthand disruption and suffering in the aftermath of the Holocaust, both in Germany and after migrating to the United States. Over time, Djurdjinovic has been able to put genetic counseling in its historical context and has come to see being a genetic counselor as "a way of taking something bad and making it good." She was reassured by the field's insistence on autonomy, a principle she stresses today as director of genetic programs at the Ferre Institute in Binghamton, New York. As Djurdjinovic puts it, "If we didn't have patient autonomy, I wouldn't be doing this."[103] Reflecting the perspective of many genetic counselors in the twenty-first century, Weil, the former director of Berkeley's genetic counseling program, asserts that sensitivity, respect, and an affirmation of active autonomy are "required to provide effective, ethical genetic counseling."[104]

Moving beyond Nondirectiveness

One of the great puzzles in the history of genetic counseling is *how* and *why* nondirectiveness attached itself so intransigently to the field, by the 1990s overshadowing an initial emphasis on autonomy and client-centeredness. Certainly Reed and many, but not all, of his colleagues expressed a commitment to nondirective approaches. However, in none of his writing does Reed actually use the term "nondirective." Instead, his contribution was the introduction of the term "client" for the counselee, a Rogerian gesture that quickly became a staple of genetic counseling. For his part, Rogers had jettisoned nondirectiveness by the mid-1940s, viewing it as ineffectual and imprecise. The new master's-degreed genetic counselors of the 1970s were too busy learning genetic science, familiarizing themselves with clinical rotations, and learning critical interviewing skills to worry much about nondirectiveness and its historical trajectory. Moreover, when they addressed issues in modern bioethics, they tended to think about autonomy, beneficence, choice, and decision making. These were the salient keywords in debates and publications generated by the genetics groups at the Hastings Center and the Kennedy Center, as well as geneticists such as Motulsky, who had fled Germany in 1939. His experience, first as a refugee passenger on a ship turned away by the Cubans in Havana Harbor and later as a survivor of internment camps in Vichy France, undergirded his awareness of potential abuses of human genetics.[105]

By the 1980s, genetic health providers accepted it as a kind of historical truism that nondirectiveness was a fundamental tenet of genetic counseling. Yet, nondirectiveness was not at the crux of debates over the ethics of genetic counseling and the role of the genetic counselor in the 1970s. Indeed, it was not until the 1980s that the static noun "nondirectiveness" captivated the field. Searches with full-text bibliographic databases demonstrate that "nondirective" and "genetic counseling" did not begin to appear together in scholarly works in health and psychology until the early 1970s, and that "nondirectiveness" had barely entered the lexicon by the 1980s. During this decade, genetic counselors, now with some professional standing and critical mass, began to assess the legacies of eugenics and abuses such as human experimentation, and they struggled with what this history might mean to their field. In so doing, genetic counselors engaged in a form of professional identity construction by producing a flood of papers and discussions about what could be seen as a phantom term. The ethical weight of the issues entangled with terminating pregnancies based on screening or diagnostic tests compelled genetic counselors to articulate a working language to navigate the difficult moral terrain of genetic counseling. It seems that the sometimes contentious debates about nondirectiveness have been cathartic and beneficial for genetic counselors and helped to chart a more emboldened path to twenty-first-century genetic screening and testing. Increasingly, however, many genetic counselors hope that nondirectiveness recedes, preferring to engage in conversations about the strengths and limits of autonomy, consent, and a broader menu of bioethical principles.[106]

Using the term "client" encouraged a more psychological, less clinical approach to the person being counseled. However, using "client" could also diminish the medical authority of the genetic counselor. Among the early wave of contemporary genetic counselors that graduated starting in the 1970s, employing the term "client" allowed them to approach the person being counseled beyond a strictly medical paradigm and was beneficial insofar as it provided great latitude for psychosocial interaction.

The problematic bind is that contemporary genetic counselors exalted client-centeredness as understandings of the patient were changing, from passive object to active subject. Today, it seems crucial that genetic counselors, whatever terminology they employ, manage to retain the psychological benefits of being client centered while fully incorporating the bioethical dimensions of beneficence, consent, and patient autonomy.

CHAPTER SEVEN

Prenatal Diagnosis

The Handmaiden of Contemporary Genetic Counseling

In 1975, Virginia Corson received her master's degree in genetic counseling. A graduate of Sarah Lawrence College, Corson identified with the emerging profile of genetic counseling students, no longer mainly housewives interested in meaningful part-time work, but now younger students training for full-time careers. During the summer after graduation, Corson searched industriously for a job, dispatching letters to the more than 50 medical genetics centers listed in the March of Dimes directory. Soon she received a call from Haig Kazazian, a pediatric geneticist at the Johns Hopkins Hospital, who offered her a position as that institution's first genetic counselor.[1]

Corson, who today is a senior genetic counselor at the Johns Hopkins Hospital, arrived during a meteoric expansion in clinical genetic services. Johns Hopkins already had a long tradition of research in human genetics, starting with the founding of a medical genetics group in 1953 and the addition to the staff four years later of Victor McKusick, who organized the Division of Medical Genetics.[2] McKusick pursued research on heritable disorders of connective tissues, notably Marfan syndrome, osteogenesis imperfecta, and Hurler syndrome, and over time became well known for his leadership in human genetic mapping through linkage studies and assembling the master list of genetic diseases in *Mendelian Inheritance in Man*, first print based and now the chief electronic database for an ever-growing list of disorders with identified genetic etiologies and influences. In addition to McKusick, Hopkins' faculty included the pediatricians Barton Childs, Robert Cooke, Barbara Migeon, and Neil Holtzman, as well as the obstetrician and future dynamo of in vitro fertilization Howard Jones.[3]

Corson joined pediatric genetics upon her arrival and enjoyed getting to know the children and families who visited the clinic, many on a regular basis. Part of Corson's responsibility involved working on the

Tay-Sachs disease prevention program, a collaborative undertaking involving physicians, scientists, and patients in Baltimore and Washington, D.C. The first of its kind in the country, the Tay-Sachs program demonstrated the potential benefits and ethical conundrums associated with the public health genetic screening of designated populations who are at substantially higher risk for this disease, specifically Ashkenazi Jews. Tay-Sachs disease is an autosomal recessive lipid storage disorder with a prognosis so poor that children born with this condition typically begin to exhibit symptoms at approximately 6 months and do not live beyond 5 years of age.[4] Likely because of his expertise and interest in expanding prenatal diagnosis to test for hemoglobinopathies such as sickle cell anemia and thalassemias, Kazazian took over the directorship of this program in the mid-1970s after its originator, Michael Kaback, moved to California.[5] Corson worked well with Kazazian, and when he was appointed director of the Prenatal Diagnostic Center (PDC) in 1978, Corson moved with him.[6] Three decades later, Kazazian describes Corson as "terrific" and a tremendous asset in terms of case management and patient communication. Corson still works at the Johns Hopkins Hospital, today as one of four genetic counselors in the prenatal unit, out of more than 20 genetic counselors employed throughout the hospital in primary and specialty clinics.[7]

Over the course of her career, Corson has been at the cutting edge of new genetic tests and technologies and has adjusted her genetic counseling practice in tandem with these developments. Since the 1970s, prenatal testing has grown from amniocentesis to include chorionic villus sampling, alpha-fetoprotein (AFP) screening, serum integrated screening, quad marker screening, and nuchal translucency, as well as a lengthening list of tests for specific conditions and susceptibilities.[8] Reflecting on her career, Corson notes, "What has occurred over the years is the great expansion in reproductive options. These changes have been both interesting and exciting but also have made my job more complicated. We've progressed from offering only amniocentesis to an evolving a la carte menu that makes sessions more lengthy and more challenging."[9]

Corson's career illustrates the intertwined relationship between genetic counseling and prenatal diagnosis, both of which emerged in earnest in the 1970s.[10] The rapid rise of amniocentesis during that decade fostered growth and professionalization, helping to create clinical arenas for employment where genetic counselors gradually managed to carve out considerable professional autonomy. Luba Djurdjinovic, like Corson a Sarah Lawrence graduate, remembers the 1970s as an exciting and formative decade in which newly minted genetic counselors demonstrated the value of their skill set, often taking the lead in setting up amniocentesis services and creating a professional milieu for "thoughtful discussions with women and their partners about this testing."[11] Through this process, genetic counselors shaped their own professional identity and learned experientially how to implement the psychosocial techniques emphasized during their training. As Corson attests, "One of the big attractions of prenatal is the independence and ability to have many genetic counseling sessions, helping with decision-making."[12]

Nevertheless, the marriage of prenatal testing and genetic counseling had its share of serious problems. The entry of genetic counseling graduates into the prenatal arena underscored and circumscribed the field as primarily a women's field, concerned with pregnancy, reproduction, children, and families. Furthermore, it placed genetic counselors in the tricky position of navigating the countervailing currents of upholding reproductive autonomy, including abortion, and supporting people with disabilities, some of whom have spoken out the most loudly against the neo-eugenic orientation of genetic screening and testing.[13] In the twenty-first century genetic counseling encompasses much more than prenatal diagnosis or pediatric genetics, having been integrated into specialties ranging from neurology to oncology. Yet the consolidation of the field was inextricably linked to the launching and routinization of prenatal testing.[14] The Johns Hopkins Hospital stood at the forefront of this development, introducing amniocentesis to American women, helping to coordinate the first systematic study of the procedure's safety, and encapsulating the complex relationship between genetic technologies and reproductive choices.

Launching Amniocentesis in America

In 1968 a network of well-seasoned geneticists, pediatricians, obstetricians, and allied health care professionals came together to launch a novel clinical service.[15] Announcing that "the scientific capability now exists for the detection of defect or the reassurance of normality in a sizable number of genetic abnormalities," this group of MDs and PhDs at the Johns Hopkins Hospital were frustrated that there was no "significant service program" in the country that could offer prenatal diagnostic services.[16] To fill this gap, they drafted a proposal to form such a center and submitted it to the March of Dimes (formerly called the National Foundation for Infantile Paralysis), which looked favorably on their plan, awarding them a substantial grant the following year. Following its triumphant sponsorship of the polio vaccine trials in the 1950s, by the 1960s the March of Dimes was channeling its considerable resources into what were broadly known as birth defects and congenital malformations.[17] Aware of the emergent technology of amniocentesis and its potential for genetic diagnosis, but also treading cautiously on the intrinsic ethical issues related to abortion, the March of Dimes provided the Johns Hopkins Hospital's PDC with a $100,000 grant.[18] Opened in 1969, the center offered the most up-to-date genetic and reproductive technologies in the country.[19]

From its inception the PDC was interdisciplinary, involving 22 faculty members from various clinical and basic science departments.[20] Administratively housed in the Department of Gynecology and Obstetrics, the center was physically located on the first floor of the Woman's Clinic. Its multitude of services required the contribution of several departments and laboratories. For example, the Department of Obstetrics and Gynecology ran the cytogenetics laboratory, the Pediatrics Department diagnosed metabolic disorders, and clinical services were provided at the Johns Hopkins Hospital and beyond at satellite clinics run jointly with the state of Maryland's Division of Preventive Health. One of the areas of ongoing negotiation among these various units were appropriate fees for procedures and cytogenetic studies, which, as a general policy, were waived for low-income patients. Progressively minded clinicians who served a large African American clientele, the PDC affiliates declared that no discrimination "on the basis of economic or racial characteristics" would be tolerated.[21]

The PDC was capitalizing on several converging streams in reproductive and laboratory medicine. Foremost among these was the procedure of amniocentesis, which is the technique of extracting amniotic fluid transabdominally through a hollow catheter or needle. The word combines the Greek *amnion* for "lamb" with *kentesis* for "puncture," imbuing it with the religious and moral watermark of the sacrificial lamb.[22] German physicians developed the procedure in the 1880s to relieve pressure brought on by hydramnios, or excessive fluid in the amniotic sac.[23] Adumbrating the development of the much safer technology of ultrasonography several decades later, in 1916 a Chicago physician was the first to diagnose a fetal defect, anencephaly, using X-ray imaging.[24] In the 1930s, physicians used amniocentesis in conjunction with emerging radiographic techniques to inject contrast media in order to assess the fetus and locate the placenta.[25] By the 1950s, obstetricians started to use amniocentesis to extract bilirubin to test maternal-fetal Rh compatibility.

Along another track, twentieth-century developments in biochemical genetics and cytogenetics were enabling the identification and diagnosis of metabolic conditions and chromosomal anomalies. In the early 1900s, scientists rediscovered the theories of Gregor Mendel, who had proposed in 1865 that hereditary material was transmitted from parent to offspring according to patterns of segregation and independent assortment. Several years later the British pediatrician Archibald Garrod postulated the "inborn errors of metabolism" theory of genetic disease based on his diagnosis of the condition alkaptonuria, or black urine, among several families in the London area.[26] By 1910, the terms "genes" and "genetics" had been coined, and during that same decade teams of scientists, led most prominently by Thomas Hunt Morgan at Columbia University, pursued gene linkage studies on *Drosophila* in the famous *Drosophila* "fly rooms."[27] In the 1950s, two scientists confirmed that humans have 46 chromosomes, another group discovered that sex chromosomes could be distinguished by the presence or absence of sex chromatin or Barr body, Jérôme Lejeune identified trisomy 21 as the cause of Down syndrome, and various scientists succeeded in growing fetal cells in culture and also were able to perform various analyses on uncultured cells.

By the mid-twentieth century a growing network of geneticists at institutions scattered across the world were identifying chromosomal anomalies, including trisomies, translocations, and sex chromosomal disorders, with new imaging technologies and laboratory techniques.[28] In 1959, for example, physicians in Denmark performed the first prenatal diagnosis for sex chromatin on a pregnant woman who knew from previous births that she was a carrier of hemophilia, an X-linked trait. Utilizing amniocentesis, the clinicians determined that she was carrying a female child, information that resulted in her choosing to continue the pregnancy.[29]

Along with ultrasonography, which localizes the placental and fetal head position, these converging developments set the stage for chromosomal and biochemical analyses of the fetus.[30] In 1966, Mark Steele and Roy Breg determined that human amniotic cells extracted via amniocentesis could be successfully cultured and karyotyped or imaged, thereby showing the chromosomal composition of the fetus. As Steele and Breg wrote in their important article, "Human amniotic fluid has been shown to contain viable cells that will grow in culture, in sufficient quantity to be karyotyped. The epithelioid morphology of these diploid cells supports the contention that they are derived from the amnion and are therefore foetal in origin. Chromosome analysis of the foetus in utero is therefore feasible."[31]

Now, chromosomal anomalies and a subset of metabolic disorders could be detected in the fetus during pregnancy. The Johns Hopkins group was keenly aware that "tissue culture, cytogenetic and enzymatic capabilities" existed at their institution but had yet to be applied to prenatal genetic diagnosis.[32]

At Johns Hopkins and a few other institutions—including at the University of California at San Francisco Hospital, Mt. Sinai Hospital in New York City, and the University of Oregon Medical School—physicians began to offer amniocentesis to their patients.[33] Kurt Hirschhorn, then at Mt. Sinai, emphasizes the reverberating impact of amniocentesis, which by the late 1960s had become an "immediate priority" and, after culturing became easier and more reliable, "spread like wildfire across the country."[34] Following conventions of clinical research, physicians and geneticists started to publish studies of their patients who had undergone amniocentesis, asserting that the procedure was effective, carried few risks, and had great diagnostic potential. For example, after performing 162 amniocenteses in the late 1960s, Henry Nadler and Albert Gerbie of Children's Memorial Hospital in Chicago reported in the *New England Journal of Medicine* that they had achieved "successful cultivation of amniotic-fluid cells" in 97 percent of cases and that amniocentesis carried "minimal risks to mother and fetus."[35] These early adapters understood the potential and implications of prenatal diagnosis and positioned it at the center of the evolving field of genetic counseling. Nadler and Gerbie observed that "the ability to detect a genetic defect in the fetus not only gives new precision to genetic counseling but also creates many moral, legal, and medical problems."[36] Similarly, reflecting on their experiences at UCSF, Charles Epstein and colleagues wrote in the *American Journal of Human Genetics* that "the ability to assess the chromosomal and metabolic status of the fetus by analysis of cultured amniotic fluid cells has converted genetic counseling from a relatively passive to an active aspect of medical practice."[37]

Apace with these medical and technological developments, many

Fig. 1—Amniotic-fluid cells.
One non-viable cell has taken up trypan-blue stain; two viable cells remain unstained.

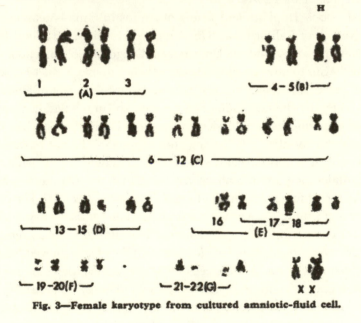

Fig. 3—Female karyotype from cultured amniotic-fluid cell.

Images included in Steele and Breg's influential 1966 study, published in the *Lancet,* which demonstrated that cells extracted through amniocentesis could be successfully karyotyped for cytogenetic analysis. This finding helped to pave the way for the use of amniocentesis for chromosomal and genetic diagnosis in the 1970s. *Source*: Reprinted from *The Lancet* 287, Mark W. Steele and W. Roy Breg Jr., "Chromosome Analysis of Human Amniotic-Fluid Cells," 383–85, Copyright 1966, with permission from Elsevier.

Americans were learning about the advent of prenatal diagnosis through the media, which sought to explain the technical aspects of these new procedures and considered their consequences for health professionals, patients, and society at large.[38] As *New York Times* health columnist Jane Brody wrote in 1971, "Prenatal diagnosis—the detection of birth defects at a time when the fetus can still be aborted—is drastically changing the nature and potential of genetic counseling throughout the country."[39] Notably, amniocentesis was performed during the second trimester, ideally between the sixteenth and eighteenth menstrual weeks of pregnancy, with results available two to four weeks later.[40] Concurrently, owing largely to the feminist health movement, abortion, still illegal throughout the country, was being decriminalized. Between 1967 and 1973, for example, approximately one-third of the states legalized abortion, and the 1973 U.S. Supreme Court decision *Roe v. Wade* upheld a woman's right to terminate a pregnancy, based on her own choice during the first trimester and in consultation with a physician during the second and third trimesters.[41] Many of the physicians and scientists at the forefront of prenatal diagnosis were keen advocates of abortion rights, because they supported women's reproductive choice and because they hoped the emergent genetic technologies at their fingertips would result in disease prevention.[42]

Richard Heller, the first director of the PDC, exemplified these trends. Born in New York City in 1938, as a teenager Heller moved to Switzerland with his family, attending university and medical school in Basel and Zurich, respectively. In 1962 he took an internship at Johns Hopkins with Howard W. Jones, where he worked on the "pathology and cytogenetics of intersexuality and also pediatric gynecologic problems in general." Soon after, Heller became a resident at Baltimore City Hospitals (which were affiliated with the Johns Hopkins Hospital) and then joined the U.S. Army and was stationed in Aberdeen, Maryland, where he practiced inpatient and outpatient pediatrics.[43] Heller returned to the Pediatrics Department at Hopkins in 1969 and quickly took a leadership position in the PDC. In this capacity, Heller helped to coordinate an interdisciplinary network of physicians and scientists connected to the PDC, sometimes running interference among strong personalities and differing departmental milieus, and acting as the public face of prenatal diagnosis at the Johns Hopkins Hospital. Embracing this role, Heller expended a great deal of time and energy responding to Right to Life groups in the Baltimore area that were denouncing proposed legisla-

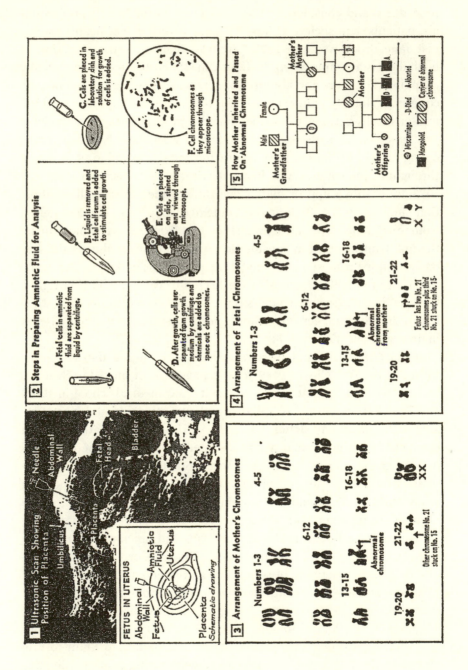

1 Ultrasonic Scan Showing Position of Placenta

Needle · Abdominal Wall · Umbilical · Placenta · Fetal Head · Bladder

FETUS IN UTERUS
Abdominal Wall · Fetus · Amniotic Fluid · Uterus · Placenta
Schematic drawing

2 Steps in Preparing Amniotic Fluid for Analysis

A. Fetal cells in amniotic fluid are separated from liquid by centrifuge.

B. Liquid is removed and fetal calf serum is added to stimulate cell growth.

C. Cells are placed in laboratory dish and solution for growth of cells is added.

D. After growth, cells are separated from growth medium by centrifuge and chemicals are added to space out chromosomes.

E. Cells are placed on slide, stained and viewed through microscope.

F. Cell chromosomes as they appear through microscope.

3 Arrangement of Mother's Chromosomes

Numbers 1-3 · 4-5 · 6-12 · 13-15 · 16-18 · 19-20 · 21-22

Abnormal chromosome

Other chromosome No. 21 stuck on No. 15

4 Arrangement of Fetal Chromosomes

Numbers 1-3 · 4-5 · 6-12 · 13-15 · 16-18 · 19-20 · 21-22

Abnormal chromosome from mother

Fetus has two No. 21 chromosomes plus third No. 21 stuck on No. 15.

X Y

5 How Mother Inherited and Passed On Abnormal Chromosome

Male · Female

Mother's Grandfather · Mother's Mother · Mother · Mother's Offspring

Miscarriage · D-Died · A-Aborted · Mongoloid · Carrier of abnormal chromosome

tion to legalize abortion. In 1971, for example, Heller countered critics of amniocentesis who had penned a letter to a local newspaper, "The purpose of the Prenatal Birth Defects Prevention Center is to assist those couples at high risk for birth defects to bear healthy children. Our center does not attempt to influence parental decisions concerning the continuation of pregnancy following diagnostic studies."[44]

As amniocentesis began, there were concerns in the medical community about its safety. In 1968, the World Health Organization issued a report on genetic counseling that warned that amniocentesis was hazardous and "involves a serious risk" to the development of the fetus.[45] With more and more physicians using amniocentesis and the legalization of abortion likely to accelerate its utilization, the federal government decided to support a systematic scientific study of the safety of amniocentesis. In the early 1970s, the National Institutes of Health's National Institute of Child Health and Human Development (NICHD) funded nine prenatal clinics affiliated with prominent hospitals across the country to create an amniocentesis registry to conduct the first blind control study of amniocentesis. This collaborative group called itself the NICHD National Registry for Amniocentesis Study Group. From July 1, 1971, to June 30, 1973, these centers—which included the Johns Hopkins Hospital's PDC, UCSF, University of California at Los Angeles, Children's Memorial Hospital in Chicago, Massachusetts General Hospital, University of Michigan, University of Pennsylvania, Yale University, and Mt. Sinai—studied the experiences of 1,040 women who had undergone amniocentesis. These study patients were matched, with some incongruities, to 992 controls with similar racial, age, and income characteristics and who, for various reasons, had not undergone amniocentesis.[46] Reviewing maternal complications, fetal loss, newborn evaluations, and the status of 1-year-old infants of mothers in the study sample, the group concluded that "results of this study provide evidence, for the first time, that midtrimester amniocentesis, despite its potential risks, is a safe procedure."[47] Specifically, the group calculated a 2 percent rate of complications such as vaginal bleeding or amniotic fluid leakage but no statistically significant fetal loss.

The data collected for this study captured the general patterns associated with amniocentesis during its initial decade. Women sought out, or were referred for, amniocentesis because of advanced maternal age (35 years or older), a history of miscarriage or stillbirth, existing children with genetic conditions, interest in knowing the recurrence risk in a second or third child (most commonly Down syndrome), family history of an X-linked genetic disorder, or family history of rare but diagnosable genetic disorders. Once amniocentesis was performed, a woman's amniotic cells were cultured and tested for cytogenetic indications, which involved counting and labeling chromosomes. The prinicipal aim was to identify trisomies and sex chromosome anomalies and to determine the sex of the fetus (XX or XY) to evaluate the risk of developing or being a carrier of an X-linked disorder such as hemophilia or certain forms of muscular dystrophy. In addition, researchers were finding that by extracting and isolating specific proteins, metabolites, and enzymes from the amniotic fluid or from cultured cells, they could prenatally diagnose rare conditions such as Hurler syndrome, myotonic dystrophy, and Fabry disease.[48]

The NICHD study's validating results propelled the growth of amniocentesis in America. While only 300 amniocentesis procedures were performed from 1967 to 1971, that figure jumped to 3,000 in 1974, 15,000 in 1978, and by 1990, more than 200,000 procedures were done.[49] During the early 1970s, clinics and hospitals across the country added prenatal diagnostic services, such as Brigham and Women's Hospital in Boston, which began to provide amniocentesis in 1974.[50] At UCSF, amniocenteses rose from less than 100 per year in the early 1970s to nearly 1,200 by late in that decade.[51] Coinciding with the growth of genetic counseling programs in the 1970s, amniocentesis became a standard part of prenatal care, above all for women over 35 and those with family histories of various genetic conditions.[52] Emulating these trends, the PDC's patient load rose rapidly during the 1970s, from 12 patients in 1969 to 25 in 1970, 102 in 1973, 201 in 1975, and more than 300 in 1976.[53]

Who came to the PDC and for what reasons? In one calendar year for which there is complete data, 1976, the PDC saw a total of 325 patients. Of these, 281 were white, 39 were black, and 5 were classified as racially "other." In addition, 218 were Protestant, 72 Catholic, 24 Jewish, and 11

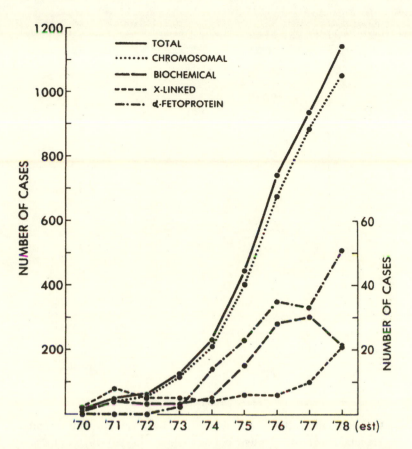

This graph illustrates the rapid rise of prenatal diagnostic services, above all amniocentesis, during the 1970s. These data were collected at the University of California at San Francisco genetics clinic, the central training site for, and eventual employer of, many of the University of California at Berkeley's genetic counseling students. *Source*: Mitchell S. Golbus et al., "Prenatal Genetic Diagnosis in 3000 Amniocenteses," *New England Journal of Medicine* 300, no. 4 (1979): 158. © Massachusetts Medical Society.

listed as other. The vast majority hailed from the region: 174 were from Maryland, outside of Baltimore, with 80 from metropolitan Baltimore, and the remainder from neighboring states, plus a handful of international patients. Most (174) were seen for cytogenetic indications, and of these, 158 because of advanced maternal age. Additional indications included a previous child with aneuploidy (usually trisomy 21), fetal sex

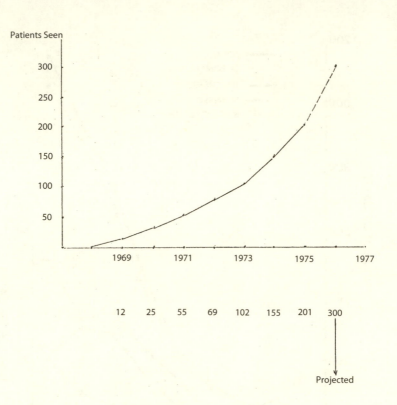

Patients Seen

| 12 | 25 | 55 | 69 | 102 | 155 | 201 | 300 |

Projected

Number of patients seen 1/1/76 to 6/30/76 = 148

This graph illustrates the rapid growth of the Johns Hopkins Hospital Prenatal Diagnostic Center from its inception in 1969-1970 to 1976. *Source*: Howard W. Jones Jr. Papers, Prenatal Diagnostic Center, Box 504035, The Alan Mason Chesney Medical Archives of the Johns Hopkins Medical Institutions.

determination, neural tube defects, metabolic disorders including Tay-Sachs disease and Hurler disease, and finally sickle cell anemia and thalassemias.[54]

Initially the PDC saw some women for nongenetic concerns, including infectious conditions and the potential side effects of pharmaceuticals or chemical exposure. For example, in 1973 a woman with "clinically identifiable Rubella" came to the clinic, as did a pregnant patient being treated for tuberculosis with the drugs Rifanpin and INH, as well as a woman of advanced maternal age (42 years old) who tested positive for syphilis.[55] Over time, however, the consultations became more explicitly focused on genetic conditions, and these always constituted the majority. For example, in January 1973, a local resident presented herself "for the fourth time" because she was pregnant and had lost a brother to Duchenne muscular dystrophy, a sex-linked disorder. As a woman she would not develop the condition, but this patient was concerned that she might be a carrier and did not want to give birth to an affected son. She already had experienced pregnancy loss and desired amniocentesis for sex selection. Records indicate that the amniocentesis was performed, although we do not know the final outcome.[56] Later that year the PDC saw a pregnant woman who was the mother of a child with mental retardation and had a history of miscarriage.[57] In addition, one of her sisters had developmental disabilities. Investigation of her family history led the PDC to collaborate with colleagues at institutions in Mississippi, where many of this patient's relatives lived. Ultimately they determined she was part of an "extensive kindred" with a translocation between chromosomes 9 and 13, a rare disorder that was "widespread throughout the family."[58]

The following year a woman in her 25th week of pregnancy was referred by her local obstetrician in Pennsylvania and sought out the PDC's services because the father of her child was known to have hereditary angioedema, an autosomally inherited disorder that severely affects the immune system, causing swelling, cramping, and airway blockage. Notably, Heller was concerned about the moral and legal issues involved with both prenatal diagnosis and abortions performed during the third trimester of pregnancy. Hence, he and his colleagues at the PDC adhered to the hospital's policy "not to perform an abortion beyond 20 weeks for psychosocial indications and beyond 23 weeks for fetal genetic indications."[59] In the end the parents indicated that they would deliver the baby and seek treatment if needed at Johns Hopkins.[60] That same month a Maryland resident consulted with the PDC because "she had lost one child with choanal atresia" and was taking care of "another living child with profound failure to thrive." This patient and her husband obtained genetic counseling, after which they decided to continue with the pregnancy "in the hope that their disease" was recessive and willing to accept the 25 percent risk of having an affected child.[61]

Later in the decade, in 1978, a 38-year-old woman underwent amniocentesis at the 17th week which revealed a "pericentric inversion of chromosome 6 in all of the 15 cells examined." Because the PDC team was uncertain about the possible phenotypic outcome of this inversion, they told her and her husband that there was an "increased incidence of duplications and deletions associated with inversion chromosomes so that a minor malformation due to the chromosomal anomaly was possible." Viewing the likelihood of a genetic defect as low, and because the mother had been frustrated by infertility for over 15 years before this pregnancy, the couple "decided to continue the pregnancy, and they are eagerly awaiting the birth of their child."[62] In another case that year, a 24-year-old woman wanted to know her risk of giving birth to a son with muscular dystrophy since one of her brothers had confirmed Duchenne muscular dystrophy. Because most of her male relatives were not affected, "the probability that she is a carrier was placed mathematically at roughly 10%" and, based on genetic assessments of the creatine phosphokinase (CPK) levels of her and her mother, was further decreased to a "2–3% chance that any male offspring would have Duchenne muscular dystrophy." Amniocentesis showed that she was indeed carrying a male (XY) fetus, but a fetoscopy, performed at Yale University, revealed that the fetal CPK was within normal limits. Hence, she continued the pregnancy.[63] Although many patients elected to continue with their pregnancies for medical, legal, religious, or personal reasons, many also chose to abort; typical was a 39-year-old woman whose amniocentesis, performed at 17 weeks, identified trisomy 18 in all cells examined.[64]

In 1980, a little over one decade after the founding of the PDC, prenatal diagnostic technologies were evolving and expanding, but consistent trends had been established. Amniocentesis was primed for routinization into prenatal care. Moreover, from the outset the PDC's affiliated physicians and researchers envisioned a more expansive prenatal diagnosis that included testing of AFP on amniotic fluid and on maternal blood in order to prenatally diagnose neural tube defects. They also wanted the center's work to interlock with the recently launched Tay-Sachs Screening program, the sickle cell disease program, and more generally "the metabolic and genetic research" underway at Johns Hopkins.[65] Key personnel at the center strove to expand prenatal testing in terms of laboratory work and the list of identifiable hereditary diseases, including by the mid-1970s skeletal dysplasias, hemoglobinopathies, and hyperammonemias, among others.

Whose Cost?

When the PDC began operating in 1970, it had three primary and inter-related objectives. The first was to save the state and the medical system money by reducing the number of births of "defective" children. This cost-benefit rationale was omnipresent in genetic screening and testing programs at the onset and reflective of the emergent cost containment pressures that began to affect medical care in the 1970s and 1980s. The second was to relieve human suffering and to help patients and families. This meant both alleviating the potential physical and psychological pain of future children through ensuring their non-birth with therapeutic abortions and working with parents and families to grieve for their loss or, in some cases, prepare for the arrival of a newborn with a genetic condition or disability.

In addition to the two preceding rationales, the PDC hoped to decrease the number of genetic defects in the population. The latter goal encapsulated the eugenic overtones that characterized genetic screening and testing in the United States. Hirschhorn, then president of the American Society of Human Genetics, captured this perspective when he described the aims of prenatal testing in 1969: "If all unborn children who were affected by or were carriers of the genes for cystic fibrosis and sickle cell anemia—two of the most common inherited diseases in this country—were identified and aborted, 14 million abortions over the next 40 years could eradicate these genes from the population."[66] A gentle version of the eugenic rationale was included in an early abstract for the center: "Our objective is healthy children reared in a loving home," a message that the PDC's first director, Richard Heller, regularly delivered in lectures to local medical and community groups.[67] For example, at a seminar cosponsored by Peninsula General Hospital and the Lower Shore Chapter of the March of Dimes, he told more than 100 physicians and nurses interested in learning more about these new prenatal technologies that "there is growing urgency among the public to produce perfect children," and he suggested that "being able to test and detect parents who are likely to produce birth defective children before a baby is even conceived is genetic counseling at its finest hour."[68] Heller continued to deliver this message after he left the PDC in 1977 to start his own genetic counseling clinic at the Greater Baltimore Medical Center. His career, however, was cut short when he died unexpectedly in 1982 at the age of 43 while jogging.[69]

Most often, the expansion of prenatal diagnosis was justified by arguments about cost-effectiveness.[70] Foremost among the reasons for the founding of the PDC was the following sentiment: "The guarantee of a normal child as contrasted with the possibility of the birth of a defective child in economic terms is well worth the small amount of money that a family might pay for this information." Indeed, the proposal submitted to the March of Dimes included the example of the cost of a "mongoloid child" as "on the order of $3,000 per year additional cost." Prevention was paramount: "The family would be spared this sort of financial burden by proper genetic counseling based on confirmed evidence of the absence or presence of defect."[71]

Early implementers of genetic testing and screening often justified their programs for Tay-Sachs disease or sickle cell anemia with arguments about cost savings. For example, in 1973 Neil Holtzman, then director of the Johns Hopkins University's Pediatric Genetics Unit, wrote to Benjamin White, the head of Maryland's Division of Preventive Medicine, to thank him for generous financial support for the Tay-Sachs disease screening program, in which Kaback, Kazazian, and Corson all participated. During its first phase, the program screened more than 10,000 residents, identifying 11 couples in which both husband and wife were carriers of this recessive gene. Of these 11 couples, four were pregnant, one of whose fetus was identified as homozygous for Tay-Sachs and subsequently aborted. Offering one of the main reasons why the investment in this program was worthwhile, Holtzman wrote that "the expense of diagnosing and caring for that one baby, had the program not been in effect, would have come close to the annual cost of the screening program."[72] The cost-saving refrain was heard across the country. For example, physicians at the University of Oregon justified the growth of their prenatal diagnosis program because "serious genetic disease generates an economic burden which must be borne by both the family and the public. For example, it is estimated that approximately 6,000 Down's syndrome children are born each year in the United States. The lifetime care for each institutionalized child has been estimated at between $180,000 and $250,000."[73]

These currents converged to fix advanced maternal age as the primary indicator for prenatal diagnosis. As Robert Resta has shown, age 35 was established as a threshold based on the three intertwined criteria of economic cost, medical risk/benefit, and eugenic concerns.[74] In short, the cost of potential institutionalization and care outweighed the price of the test. As for medical risks, as the NICHD study showed, the risk of miscarriage and the probability of having a child with trisomy 21 reached parity—of 1/200—at age 35 (many agree that the current risks of amniocentesis are lower than this).[75] In addition, several important lawsuits for "wrongful life," in which mothers argued that their physicians neglected to offer them amniocentesis despite their advanced maternal age, an oversight that resulted in the birth of a child with severe disabilities, helped to entrench age 35 as the iconic number and imbue it with "high-risk" connotations.[76]

Researchers also employed cost-effectiveness arguments to enlarge the armamentarium of prenatal diagnosis and screening. As the PDC began its operations, its scientists and physicians saw great promise in the development of AFP testing, which analyzes the amniotic fluid not for chromosomal anomalies or metabolic disorders but for the risk of neural tube defects such as spina bifida. AFP was first identified in the 1950s, and in 1972 two researchers demonstrated that elevated levels of AFP in amniotic fluid were associated with open neural tube defects.[77]

Building on these discoveries, the PDC's John Freeman applied to the March of Dimes to support his research in the area of AFP extraction and analysis. Swiftly funded, Freeman added this procedure to the PDC's services.[78] In addition, Heller and Freeman wanted to expand this program to include testing of maternal serum, which was less invasive than fluid extracted via amniocentesis. In 1976, Heller expressed his enthusiasm about the "opportunity to assume a leadership position in the provision of a major new clinical service. It would be a damn shame to let this opportunity slip by!"[79] To bolster his case, Heller referred to a 1976 article written by veteran prenatal diagnosticians Aubrey Milunsky and Elliott Alpert of Massachusetts General Hospital, who were urging the Food and Drug Administration (FDA) to immediately license AFP testing. Invoking arguments of clinical safety and cost-effectiveness, they wrote, "We believe that the alpha-fetoprotein assay should now be available as a routine test for all samples obtained for antenatal genetic studies. The projected cost of life-time care for one infant surviving with open spina bifida is between $100,000 and $250,000. Alpha-fetoprotein testing of amniotic fluid is relatively inexpensive (<$20), and although it is nonspecific, its clinical value is now well established."[80] Based on this reasoning, in 1978 Heller and several colleagues proposed a comprehensive study of pregnancies in the state of Maryland that would involve testing *all* pregnant women for AFP determination.[81] By the mid-1980s, maternal serum alpha-fetoprotein screening was available in many pre-natal clinics across the country, and in 1985 the uptake of MSAFP rose dramatically when the American College of Obstetricians and Gynecologists Department of Professional Liability issued an alert urging that all patients be offered this screening, in part to protect physician liability.[82]

Choices in Context

When the Johns Hopkins group formulated their plan for the creation of the PDC, having a genetic counselor on staff was an integral component. As they wrote, "The clinical services which are to be provided will be those of obstetrical counseling through a genetics counselor."[83] Before Corson was hired in 1975, physicians, including Heller, and Kazazian, who was closely affiliated with the unit, offered genetic counseling. Soon after the PDC opened, local papers informed readers about newly available services. In November 1974, for example, the *Advertiser*, published in Salisbury, Maryland, announced that residents could "Dial the Doctor" to listen to a "fact filled" message on genetic counseling and the services offered at the PDC.[84] For his part, Kazazian remembers offering genetic counseling according to general principles of good bedside manner, drawing from "the training I had in medical school in terms of dealing with people, in terms of trying to be sympathetic, and yet presenting the information and trying to be objective about it, trying to be comforting."[85] Even so, Kazazian relied on the traditional physician's model of information delivery, and perhaps for this reason he recognized the need for someone like Corson, trained psychosocially to talk with, rather than to, the patient.

Throughout the country, newly minted counselors like Corson helped to establish and expand prenatal genetic testing. Some, like Djurdjinovic, who graduated from Sarah Lawrence in 1978, had the opportunity for a clinical rotation at Mt. Sinai's clinic, which served as a training site for students starting in 1975.[86] After receiving her degree, Djurdjinovic moved to upstate New York, where she was at the forefront of offering amniocentesis in a growing network of clinics devoted to women's reproductive health, including even shepherding samples of amniotic fluid to the appropriate regional laboratories. She remembers being shocked at how many physicians pushed amniocentesis. One, for example, told his patients into the late 1970s, "It is a law, my dear." Given the unabashed paternalism among many of the obstetricians and family physicians she encountered, Djurdjinovic felt strongly that she needed to help the physicians that she worked with understand that prenatal testing involved complex decisions and necessitated a safe environment for psychological discussion between the genetic counselor and client. As she recalls, "From very early on prenatal genetic counselors had a real hand in making prenatal testing something that required a discussion," which in turn involved a psychosocial dynamic "of meeting the patient wherever she was in the process."[87]

Elsa Reich, also a Sarah Lawrence alumna, played a critical role in starting the amniocentesis clinic at New York University in 1975. Working with the director of medical genetics and several members of the obstetrics department, Reich created a protocol for women undergoing prenatal testing. Notably, she had never counseled anyone for amniocentesis as a Sarah Lawrence student, so once she began to offer prenatal counseling she literally "had to figure it out on her own." Reich vividly recalls the primitiveness of the technology, in comparison with what is available today. The ultrasound images were snowy, testing was available only for chromosomal anomalies, and AFP testing for neural tube defects was barely on the horizon. Offering services three days a week at the beginning, Reich recollects a small but steady stream of patients, seeking out amniocentesis because they were 35 or older, whose fetuses were diagnosed most commonly with trisomies 21, 18, and 13.[88]

The documents in the archives related to the PDC's first decade of operations reveal the complexity and contradictions surrounding reproductive choices in the prenatal diagnostic setting. To begin, it is difficult to glean personal experiences from highly circumscribed records such as minutes of clinical meetings and patient summaries. Yet there are some clues. For example, in 1976, a Baltimore resident underwent amniocentesis because of her advanced maternal age, and her fetus was diagnosed with trisomy 21. In this case, "in concert with the patient's wishes, a therapeutic abortion was performed at approximately 21 weeks of gestation." After follow-up counseling, this patient was "pleased with her experience at JHH and as an expression of her positive feelings for the nature and manner of services provided, has expressed the desire to undertake another pregnancy as soon as feasible, and to once again have the pregnancy monitored by amniocentesis."[89]

Although this patient received what she wanted, some did not. For example, in 1980, Kazazian wrote in the *Hastings Center Report* about the ethics of sex selection. In response to a position paper that argued that parents had a right to use amniocentesis to determine sex and to abort if they so chose even if the physician disagreed, Kazazian discussed the case of a 40-year-old pregnant woman and her husband who had requested midtrimester amniocentesis with the explicit aim of learning the sex of the fetus. Because of advanced maternal age, and following a counseling session, the PDC performed the amniocentesis, which revealed that the fetus was not of the "desired sex." Given the PDC's guidelines for pregnancy termination, the couple obtained an abortion at another hospital.[90] This case and circulating arguments about the importance of reproductive autonomy—even if it includes sex selection—prompted the Johns Hopkins Hospital to "liberalize" its policy. Starting in 1979, all "patients who desire sex selection are counseled," and abortions were performed if requested (although indications are that this policy later reversed). Kazazian noted that the number of such cases was exceedingly small and consisted almost exclusively of couples of "Asian extraction, frequently Indian," who "desire a male after one or more female children."[91]

As amniocentesis became routine and prenatal services were enlarged to encompass CVS and ever more sophisticated genetic screening technologies, many women began to experience pregnancy, especially until midtrimester confirmation that the fetus was "normal," as a prolonged at-risk limbo. Barbara Katz Rothman captured this development in her thoughtful and well-researched exposé of the impact of amniocentesis on reproduction, coining it "the tentative pregnancy."[92] During the first decade of amniocentesis, genetic health professionals created the circumstances for and witnessed the psychological turmoil and anxiety experienced during the phase of "tentative pregnancy." In small-scale studies of patients, they found that guilt and depression were common denominators, above all because most patients were women who felt compelled to terminate a wanted pregnancy because of an identified genetic defect.[93] In the 1970s, Epstein's group at UCSF reflected on the trials and tribulations of their first 50 patients: "A woman undergoing amniocentesis is an anxious person, and the period of waiting between the procedure and the results does not serve to diminish her anxiety. Given any unexpected difficulties, these anxieties may become very intense."[94]

Nevertheless, depression and guilt were often accompanied by relief and even contentment. Several studies demonstrated that despite the psychological difficulty of making a choice, women who terminated pregnancies because of genetic conditions felt they had made the right one.

For example, a study based on 157 questionnaires distributed to women who underwent prenatal diagnosis at the University of Alabama hospitals and clinics found that 57 percent did so because of advanced maternal age, 71 percent intended to "end the pregnancy if the test showed that you were carrying an abnormal fetus," and the most difficult part about deciding to get the procedure was "having to decide about ending the pregnancy." Nevertheless, 94 percent stated that they would use amniocentesis again if they became pregnant, and 98 percent stated they would recommend the test to others. The authors concluded, "This relatively small but ongoing study shows that the majority of women using these studies considered them a positive and reassuring experience."[95] In the early 1990s, Dorothy Wertz and John Fletcher rebuffed arguments of scholars and activists who contended that prenatal diagnosis had become a eugenic "search and destroy" mission. Based on an extensive literature review, they determined that attacks on prenatal diagnosis as heavy-handed eugenics were inaccurate and misrepresented women as victims, not agents of their own reproductive destiny. They concluded that "women exercise choice in regard to prenatal diagnosis," underscoring that "women probably have more choices about prenatal diagnosis than they do over other technologies used in birth."[96] In a study of couples who considered fetal diagnosis during pregnancy, Susan Markens and colleagues found that Mexican-origin women and their partners were agents, asserting, "Our findings support other research that has challenged the stereotypical myth of male control of decision making in Mexican-origin families in general and reproductive decision making in particular," adding that they also "challenge the radical feminist view that new reproductive technologies will be used to reinforce men's attempts to control women's bodies."[97] Their study suggests that many Mexican-origin women engage in an egalitarian form of parenting and decision making and that many draw directly from a patient autonomy / reproductive rights framework when discussing why it ultimately was the woman's choice to proceed or not with amniocentesis. More exuberantly, in her recent book on prenatal testing, which examines the screening program for ß-thalassemia in Cyprus, Ruth Schwartz Cowan has lauded prenatal diagnosis as the antithesis of eugenics.[98]

Yet even when genetic counselors are well meaning, they sometimes fail to calibrate their approaches to meet racial, ethnic, or cultural needs of their clients, a gap that can be partially explained by the profession's demographic homogeneity and paucity of minority counselors. As Carole

Browner and colleagues found in a study of Latinas considering prenatal diagnosis in California, "Counselors almost invariably followed the standard protocol regardless of clients' educational or ethnic background," even when they talked "sensitively about the often difficult issue of abortion as well as other matters such as caring for a child born with a disability."[99] They describe one clinical encounter (of the 73 they observed and ethnographically analyzed) as noticeably marred by miscommunication. In this case, a 45-year-old Mexican American woman with a history of hereditary conditions met with a genetic counselor who she felt was "too critical" and thus did not "trust her," a situation exacerbated by using a third person to translate. These kinds of miscues and misgivings suggest "that the best way for counselors to improve comprehension and win their clients' ears and understanding is to give a better listen to their clients' words."[100]

Reflections on Prenatal Diagnosis

Amniocentesis served as the technological handmaiden for the development of contemporary genetic counseling. Seymour Kessler, who directed the genetic counseling program at the University of California at Berkeley for many years, cogently writes that karyotyping and amniocentesis "opened the way for the prenatal diagnosis of chromosomal disorders," and that with these advances "modern genetic counseling had been born."[101] Alongside screening programs, above all for Tay-Sachs disease, prenatal diagnostic clinics were the primary sites for the training and formation of genetic counselors in the 1970s and 1980s. During these early decades, genetic counselors took positions in programs guided by the goals of disease or disability prevention and cost-effectiveness. In addition, genetic counseling benefited from the fresh stream of income generated by amniocentesis services and related laboratory fees. Indeed, most of the initial graduates of genetic counseling programs worked in prenatal services. Bonnie Baty, a member of the second cohort at Rutgers University and today director of the genetic counseling program at the University of Utah, distinctly remembers entering a profession that was tightly wedded to amniocentesis. Her peers Michael Begleiter and Gary Frohlich also noted the dominance of prenatal services (and to a lesser extent pediatrics) in the 1970s.[102] Robert Resta has described a genetic counselor's typical day in the 1980s as consisting almost entirely of "prenatal, offering patients amniocentesis and ultrasound," and eventually MSAFP screening.[103]

Analysis of the first decade of the PDC suggests that the reality of prenatal diagnosis encompasses a wide spectrum of personal experiences. Certainly, women make choices within broader contexts that they cannot control, and our society places unrelenting demands on mothers and families. Women are subtly and not so subtly pressured, and eugenic ideas about optimal health and well-being still underpin some aspects of the delivery and assumptions of prenatal care.[104] Yet prenatal diagnosis is just one arena where social and personal expectations weigh heavily on potential mothers and parents. Many genetic counselors see their job as helping patients navigate these pressures to arrive at whatever is the best decision for them. For example, Reich believes that reproductive choices must be based first and foremost on facts. Yet she has no "vested interest in the actual decision, only that the patient is making the right decision for her." Reich asks the patient to determine what the least acceptable outcome is for her: the risk of losing a pregnancy (given the small risk associated with the procedure of amniocentesis) versus knowing if the fetus is normal, which is the most likely outcome. Reich speaks for the vast majority of genetic counselors practicing today when she states that "choice means you have the choice not to choose to terminate a pregnancy."[105]

Conclusion

Interviewed in 2001, the geneticist Barton Childs, who spent most of his career in pediatric genetics at the Johns Hopkins University and Hospital, took stock of the dramatic changes he had witnessed over his lifetime. Cumulative scientific knowledge was demonstrating that "genetics and genomics are the basic sciences for medicine." Not only had medical education been remade from its DNA out, but earlier causal modes of thinking about disease were becoming antiquated. As Childs explained, "One of the most important elements in this transformation is the gradual recognition that linear relationships don't obtain, that causes are multiple, whether of genes or experiences of the environment, so that what we're now calling complex diseases are clearly caused by not only the specific proteins specified by genes, themselves various, but by a great variety in the kind and numbers of experiences of the environment over the lifetime."[1] Nearly a nonagenarian at the time of this interview, Childs was outlining the twenty-first-century directions of genomic medicine, which now permeates understandings of disease processes and deeply influences how we think about diagnosis and the promise of treatment. More specifically, he was heralding the arrival of epigenetics, personalized genetics, genomic sequencing, and genome-wide association studies (GWAS), all at the cutting edge of genetic science.[2]

Today's genetic counselors are cognizant that the amplification and growing ubiquity of genomic medicine present both opportunities and challenges for their profession. Yet they do not always concur on likely directions in the future. In the opinions of some, the twenty-first-century genomic revolution could lead to the eventual erosion of the niches occupied by genetic counselors over the past several decades, a scenario that Luba Djurdijnovic, trained at Sarah Lawrence and now executive director of the Ferre Institute in Binghamton, New York, finds plausible. Djurdijnovic imagines that the field "is only going to last for a period of time" and could easily be "assimilated" into the health care provided by primary and specialist physicians. Djurdijnovic, however, is sanguine rather than maudlin about it. She believes that genetic counseling has more than proven its worth, "serving as a vehicle to getting patient-centered care into clinical medicine."[3]

In contrast, other genetic counselors have a more optimistic prognosis and anticipate further growth of the profession. For example, Wendy Uhlmann, a genetic counselor at the University of Michigan and past president of the National Society of Genetic Counselors, believes that advances in genomic medicine are creating new areas for the delivery of genetic services. In her estimation, genetic counselors should and will be at the forefront of clinical genomics, particularly with respect to patient communication, education modules, and assessing how to effectively convey a lengthening list of genetic information to an array of patients and clients.[4] Catherine Reiser, director of the program at the University of Wisconsin at Madison, recognizes that genetic counseling will be buffeted by "external and internal forces" in coming years but thinks that the field has demonstrated a capacious ability to adapt, change, and flourish.[5]

Many factors will challenge the scope and reach of genetic counseling in the twenty-first century. At the top of the list is the proliferation of an ever-expanding menu of genetic tests that individuals might undergo in a doctor's office or seek out independently, usually via online companies such as 23andMe or deCODEme.com.[6] For example, the biotechnology company Sequenom, based in San Diego, recently announced a new test that analyzes fetal DNA in the mother's blood to detect Down syndrome.[7] Although this new test is being rolled out gradually, it easily could offset the utilization of riskier, more invasive procedures such as amniocentesis and chorionic villus sampling. Veteran genetic counselor Robert Resta, who has watched prenatal testing evolve over the past three decades, cautions that "if this test pans out, pregnant woman may never again see a genetic counselor," except perhaps in the minority of instances when the result is abnormal.[8]

If the advent of universally available, less invasive screening technologies poses significant challenges for today's genetic counselors, so too does the concomitant rise of commercial genetic testing, which runs the gamut from whole genome sequencing (thus far more of a curiosity for the affluent few who can afford the $5,000 to $10,000 cost) to the less expensive options of ancestry and direct-to-consumer genetic tests (usually several hundred dollars). Whatever the price tag, these tests are all growing in availability and popularity and are reshaping how Americans think about their health and identity. As the perceptive *New York Times* reporter Amy Harmon, who has written extensively about the intersections of genetics and culture in the United States, states, the wide "embrace of DNA genealogy speaks to the rising power of genetics to shape our sense of self."[9]

As of 2010, there were more than 30 DTC genetic testing companies for more than 50 health conditions. Some people want to pay out of pocket for commercial tests in order to circumvent their health insurance company or to exclude results, which might be tagged as preexisting conditions, from their medical record. Others simply want to know about their disease susceptibilities or the inheritance patterns of their biological kin. Many genetic counselors have reacted to the emergence of genetic testing for personal purchase with dismay if not disgust, expressing wariness about DTC testing, particularly its consequences for the delivery of genetic information. Many DTC companies provide uninterpreted information, accessible online and updated regularly, which consumers can peruse at their convenience. Only a handful of these companies, such as Navigenics, have incorporated genetic counseling into their overall services and make genetic counselors available to clients on an integrated, not a la carte, basis.[10] Not surprisingly, this trend concerns genetic counselors; because "DTC genetic testing does not generally involve a physician or genetic specialist, individuals who purchase DTC genetic tests are left to interpret—or misinterpret—test results."[11] Genetic health professionals have also pointed out that DTC genetic tests suffer from a noticeable lack of federal oversight and regulation.[12]

Nevertheless, some genetic counselors see positive aspects of DTC genetic testing. Beverly Yashar, who directs the genetic counseling program at the University of Michigan, acknowledges that her first response to DTC testing was, as she describes it, "very insulated and negative." Above all, she was concerned that there simply were not enough genetic counselors to handle the prospective number of patients using DTC testing. Yet as Yashar watched these technologies being rolled out, she decided that genetic counselors were being too paternalistic about them and that they needed instead to grant the "consumer or patient some power and ability to understand the results." She now views DTC testing as just one of many realms in which genetic counselors can put themselves on the map and shape genetic health and literacy in the twenty-first century.[13] Many recent graduates of genetic counseling programs will have this opportunity, given the robust expansion of jobs at for-profit companies that are at the forefront of introducing the latest genetic tests to both health care providers and consumers.[14]

Moreover, some genetic counselors view the growth of commercial genetic testing as an avenue for educating researchers and physicians and shaping the priorities of personalized genomic medicine. Karen Heller, who graduated from a small, short-lived, but demanding genetic counseling program at the University of North Texas in Denton, worked for many years in pediatric genetics at a major hospital in the Dallas area.[15] After more than 20 years in clinical settings, Heller accepted a position with a genetic testing company as a regional medical specialist. She is one of 30 such specialists at her company, most of whom are trained genetic counselors. Heller finds the position exciting and rewarding since she now "reaches patients through the doctors that she is teaching" and counsels "doctors instead of patients." Based on her experience working in the private sector, Heller believes that companies such as Myriad, which is involved in a legal battle to hold an exclusive patent on the BRCA1 and BRCA2 gene tests, get a bad rap. From Heller's perspective, entities such as Myriad and Genzyme have not limited technologies but instead taken something that is clinically valuable and made it "more accessible and more useful."[16]

As genetic counselors reflect on their roles vis-à-vis commercial testing, they also face the problem of the profession's homogeneity. The racial, ethnic, gender, and class profile forged in the 1970s remains the norm. According to the NSGC's 2010 Professional Status Survey, approximately 95 percent of genetic counselors self-identify as Caucasian, 5 percent as Asian, and 1 percent as African American. Only 5 percent are men.[17] Many of the genetic counselors I interviewed for this project want practitioners to more closely match the demographics of patients and keep apace with a multiracial and multicultural society. What is disheartening for many is that despite concerted efforts over the years, this career personality has proven exceedingly intractable. As soon as she took the helm of Sarah Lawrence's genetic counseling program in the early 1970s, for example, Joan Marks tried many times (to no avail) to secure funds for minority fellowships (and worked closely with several historically black colleges toward this goal) and to offer specialized Spanish language training. Despite similar attempts at other institutions over the years, the profession's makeup remains virtually unchanged.

According to Barbara Biesecker, director of the National Human Genome Research Institute / Johns Hopkins University genetic counseling program, the field is at a critical impasse when it comes to diversity. She suggests that one of the core reasons is that genetic counselors "have not articulated a contemporary working hypothesis for why so few under-represented minorities or men are drawn to the profession."[18] In her 2004 presidential address to the NSGC, Kelly Ormond acknowledged to a ballroom of her colleagues that "we are already well aware that our profession is under-represented in ethnic minorities as compared to our national census. The dearth of diverse genetic counselors as service providers impacts our ability to provide culturally competent care."[19] Today, as director of one of the youngest programs in the country, at Stanford University (started in 2008), Ormond hopes that being in "one of the most culturally and linguistically diverse regions of the country"—the larger San Francisco Bay area—will help her build a more multicultural student body.[20] On the east coast, Caroline Lieber, alumna of and now director of the genetic counseling program at Sarah Lawrence College, strives to change this profile through educational outreach, especially to high school and elementary schools.[21] Nancy Steinberg Warren, long affiliated with the genetic counseling program at the University of Cincinnati, has developed a cultural competence toolkit for practitioners, available online, that emphasizes health disparities and cross-cultural communication as key elements of providing patient-centered care.[22] Perhaps in the twenty-first century motivated genetic counselors, as well as a gradually more diverse student body, will be able to alter the complexion of the profession.

Despite the field's unresolved issues from the past and likely difficulties in the future, my historical research and analysis suggest that genetic counseling is here to stay. Since I started the research for this book, several new professional programs have been established, including at Stanford, neighboring California State University at Stanislaus, and the University of Alabama. Applicants to existing programs face strong competition. If state licensure continues at the current rate and consultations increasingly become recognized billable services that are reasonably reimbursed to health care providers, then genetic counseling services will become interwoven into the fabric of health care and accessed by a wider cross section of people.

Trends in genomic medicine also could make genetic counselors more visible. If so, they can use this opportunity to shake off the stereotype that they are abortion pushers. Over and over, as I presented research in progress to educated nonmedical audiences whose experiences of genetic counseling were minimal or limited to prenatal care, a common tendency was to conflate genetic counseling with pregnancy termination. As this book has shown, genetic counselors were and have been central to prenatal diagnosis. Yet, the field has encompassed much more than that. Most obviously it has influenced pediatric and adult genetics. The genetic counselors trained in the 1970s endorsed the prevention model of prenatal diagnosis, yet they also vocally advocated for reproductive autonomy—whatever the decision and outcome. If genetic counselors become even more disability-friendly and if programs such as the one started by Judith Tsipis at Brandeis University serve as imitable models, this stereotype could disappear once and for all.

Archival Materials Consulted

Alan Mason Chesney Medical Archives, The Johns Hopkins Medical
    Institutions, Baltimore, Maryland
    Howard W. Jones, Jr. Papers
    John Littlefield Papers
    Robert E. Cooke Papers
    Victor McKusick Papers
    Biographical Files
    Subject Files
American Philosophical Society Library, Philadelphia, Pennsylvania
    American Eugenics Society Records
    Curt Stern Papers
    James V. Neel Papers
Bentley Historical Library, University of Michigan, Ann Arbor, Michigan
    Adult Medical Genetics Clinic Records
    Department of Human Genetics Records
    Lee R. Dice Papers
    University of Michigan Board of Regents Records
Center for Society and Genetics, University of California at Los Angeles
    Oral History of Human Genetics Project
Dorothy Carpenter Medical Archives, Wake Forest Baptist Medical Center,
    Winston-Salem, North Carolina
    Department of Medical Genetics Records, Unprocessed Boxes
    K. Davis Papers
    Nash C. Herndon Papers
    Nash C. Herndon and William Allan Papers, Unprocessed Boxes
    William Allan Papers
March of Dimes, White Plains, New York
    Conference and Meetings Records
    Grants Records
    Medical Programs Records
    Oral History Collection
Sarah Lawrence College, Special Collections, Bronxville, New York
    Charles DeCarlo Papers
    Esther Raushenbush Papers
    Human Genetics Graduate Program Records

Rutgers University Archives, Special Collections and University Archives
   Charlotte Avers Papers
   Course Catalog Collection
   Douglass College Office of the Dean Records
University of Minnesota Archives, Minneapolis, Minnesota
   Biographical Files
   Dight Institute for Human Genetics Records
   Minnesota Human Genetics League Records
Personal archive of materials related to the University of California at Berkeley
   Genetic Counseling Program, held by Margie Goldstein, Berkeley,
   California
Personal archive of Dight Institute Inquiries, Appointment Books, 1948–1980,
   23 volumes, held by the author, Ann Arbor, Michigan

Unless otherwise noted, "Professional Training" denotes master's degree leading to board certification in genetic counseling.

Name	Professional Training	Current or Relevant Affiliation(s) (as of Oct. 2011)
Diane Baker	Sarah Lawrence College	Genetic Alliance
Bonnie Baty	Rutgers University	University of Utah Health Sciences Center
Michael Begleiter	Rutgers University	Children's Mercy Hospital, Kansas City
Robin Bennett	Sarah Lawrence College	University of Washington
Barbara Biesecker	University of Michigan	National Human Genome Research Institute / Johns Hopkins University Genetic Counseling Program
Joan Burns	University of Wisconsin	University of Wisconsin
Virginia Corson	Sarah Lawrence College	Johns Hopkins Hospital
Jessica Davis	MD, Columbia University College of Physicians and Surgeons	Weill Cornell Physicians
Luba Djurdjinovic	Sarah Lawrence College	Ferre Institute
Katy Downs	University of California at Berkeley	University of Michigan
Debra Lochner Doyle	Sarah Lawrence College	Washington State Department of Health
Janice Edwards	Sarah Lawrence College	University of South Carolina
Charles Epstein	MD, Harvard University Medical School	University of California at San Francisco
Marvin Frankel	PhD, University of Chicago	Sarah Lawrence College
Gary Frohlich	Rutgers University	Genzyme Therapeutics
Robin Grubs	University of Pittsburgh (also PhD)	University of Pittsburgh
Bryan D. Hall	MD, University of Louisville Medical School	University of California at San Francisco

Audrey Heimler	Sarah Lawrence College	Long Island Jewish Medical Center
Karen Heller	University of North Texas	Myriad Genetics
Kurt Hirschhorn	MD, New York University	Mt. Sinai Hospital
Annette Kennedy	PsyD, Massachusetts School of Professional Psychology	Brandeis University
Seymour Kessler	PhD, Columbia University; PhD, Wright Institute	University of California at Berkeley
Bonnie LeRoy	Sarah Lawrence College	University of Minnesota
Caroline Lieber	Sarah Lawrence College	Sarah Lawrence College
Dorene Markel	University of Michigan	University of Michigan
Joan Marks	MSW, Simmons College	Sarah Lawrence College
Jacquelyn Mattfeld	PhD, Yale University	Sarah Lawrence College
Anne Matthews	University of Colorado Health Sciences	Case Western Reserve University
Arno Motulsky	MD, ScD, Yale University	University of Washington
Carol Norem	University of California at Berkeley	Kaiser Permanente
Kelly Ormond	Northwestern University	Stanford University
Roberta Palmour	PhD, University of Texas at Austin	McGill University (Montreal)
June Peters	Rutgers University	National Cancer Institute, National Institutes of Health
Lucille Poskanzer	University of California at Berkeley	Oakland Children's Hospital
Diana Puñales-Morejon	Sarah Lawrence College; PhD, Columbia University	Sarah Lawrence College, CUNY Psychological Center
Kimberly Quaid	PhD, Johns Hopkins University	Indiana University School of Medicine
Elsa Reich	Sarah Lawrence College	New York University
Catherine Reiser	University of Wisconsin	University of Wisconsin
Roberta Resta	University of California at Irvine	Swedish Medical Center, Seattle
Marian Rivas	PhD, Indiana University	Rutgers University
Judith Tsipis	PhD, Massachusetts Institute of Technology	Brandeis University
Wendy Uhlmann	University of Michigan	University of Michigan
Ann Walker	University of California at Irvine	University of California at Irvine
Nancy Steinberg Warren	Sarah Lawrence College	University of Cincinnati

Jon Weil	PhD, University of California at Davis; PhD, Wright Institute	University of California at Berkeley
Beverly Yashar	PhD, University of North Carolina at Chapel Hill	University of Michigan

Master's Degree Genetic Counseling Programs in North America

This list includes programs established since 1969 and is accurate as of May 2012. For an international list of programs see http://tagc.med.sc.edu/education.asp. The establishment date refers to when each program was accredited per the criteria in force at the time (some accreditations here are provisional) or when the program accepted its first student or cohort, with the assumption that the inaugural cohort graduated 18 to 24 months after establishment of the program.

Institution	Location	Established	Suspended
Sarah Lawrence College	Bronxville, NY	1969	
Rutgers University (Douglass College)	Rutgers, NJ	1971	1980
University of Pittsburgh	Pittsburgh, PA	1971	
University of Colorado Health Sciences Center	Denver, CO	1971	
University of California, Berkeley	Berkeley, CA	1973	2004
University of California, Irvine	Orange, CA	1973	
State University of New York, Stony Brook	Stony Brook, NY	1974	1979
University of Wisconsin	Madison, WI	1976	
Howard University	Washington, DC	1976	
University of North Texas	Denton, TX	~1978	1980
University of Michigan	Ann Arbor, MI	1980	
University of Cincinnati	Cincinnati, OH	1981	
University of South Carolina	Columbia, SC	1985	
University of Texas Health Science Center at Houston	Houston, TX	1989	
University of Minnesota	Minneapolis, MN	1989	
Northwestern University	Chicago, IL	1990	
Virginia Commonwealth University (Medical College of Virginia)	Richmond, VA	1990	
Indiana University	Indianapolis, IN	1991	
Brandeis University	Waltham, MA	1992	

California State University, Northridge	Northridge, CA	1995	2008
Arcadia University	Glenside, PA	1995	
University of Maryland	Baltimore, MD	1995	
Mt. Sinai School of Medicine	New York, NY	1996	
Johns Hopkins University / National Human Genome Research Institute	Baltimore and Bethesda, MD	1996	
University of British Columbia	Vancouver, BC, Canada	1996	
University of Arizona	Tucson, AZ	~1998	2005
Case Western Reserve University	Cleveland, OH	1998	
University of Toronto	Toronto, ON, Canada	1998	
University of North Carolina	Greensboro, NC	2000	
Wayne State University School of Medicine	Detroit, MI	2001	
Boston University School of Medicine	Boston, MA	2004	
University of Arkansas for Medical Sciences	Little Rock, AK	2006	
California State University, Stanislaus	San Francisco Bay area, CA	2008	
University of Utah	Salt Lake City, UT	2008	
Stanford University	Palo Alto, CA	2008	

PROGRAMS WITH PROVISIONAL ACCREDITATION FROM THE ABGC

Institution	Location	Established	Suspended
University of Oklahoma	Oklahoma City, OK	2003	
University of Alabama	Birmingham, AL	2010	
Long Island University	Brookville, NY	2010	
Emory University School of Medicine	Atlanta, GA	2011	

INTRODUCTION

1. *2010 Professional Status Survey: Salary and Benefits* (Chicago: National Society of Genetic Counselors, 2010), 18–19, available at www.nsgc.org.

2. Between 2007 and 2011, I interviewed 46 individuals either by phone or in person for 60 to 120 minutes. Interviewees included directors of over half of the genetic counseling master's degree programs in existence from 1969 to 2012. See appendix B for a complete list of interviewees.

CHAPTER ONE: *History*

1. Judith Tsipis, interviewed by the author, August 5, 2010.

2. Robert Cook-Deegan, *The Gene Wars: Science, Politics, and the Human Genome* (New York: W. W. Norton, 1994).

3. Tsipis, interview.

4. Ibid.

5. National Tay-Sachs and Allied Diseases Association, Inc., "What Is Canavan Disease?," (Boston, 1999).

6. "Macrocephaly," in Helen V. Firth, Jane A. Hurst, and Judith G. Hall, *Oxford Desk Reference: Clinical Genetics* (Oxford: Oxford University Press, 2005), 162–63.

7. Tsipis, interview.

8. Ibid.; Obituaries, MIT News, January 28, 1998, available at http://mit.edu/newsoffice/1998/obits-0128.html.

9. Wendy R. Uhlmann, Jane L. Schuette, and Beverly M. Yashar, *A Guide to Genetic Counseling*, 2nd ed. (Hoboken, NJ: Wiley-Blackwell, 2009).

10. The most up-to-date list of accredited programs can be viewed at the website of the American Board of Genetic Counseling, www.abgc.org.

11. *2010 Professional Status Survey: Work Environment* (Chicago: National Society of Genetic Counselors, 2010), available at www.nsgc.org.

12. Ibid.

13. Regina H. Kenen, "Opportunities and Impediments for a Consolidating and Expanding Profession: Genetic Counseling in the United States," *Social Science & Medicine* 45, no. 9 (1997): 1377–86; also see www.abgc.net/ABGC/AmericanBoard ofGeneticCounselors.asp and www.nsgc.org.

14. *2010 Professional Status Survey*; Dorothy C. Wertz and John C. Fletcher, "Communicating Genetic Risks," *Science, Technology, & Human Values* 12, no. 3/4 (1987): 60–66; Marie-Louise Lubs, "Does Genetic Counseling Influence Risk Atti-

tudes and Decision Making?," *Birth Defects: Original Article Series* 15, no. 5C (1979): 355–67.

15. Lori B. Andrews, *Future Perfect: Confronting Decisions about Genetics* (New York: Columbia University Press, 2001).

16. Robert Marion, *Genetic Rounds: A Doctor's Encounters in the Field That Revolutionized Medicine* (New York: Kaplan, 2009).

17. For the most up-to-date information on genetic testing, see www.ncbi.nlm .nih.gov/sites/GeneTests/?db=GeneTests.

18. For these and many more statistics on genetic conditions, see www.kumc.edu/ gec/prof/prevalence.html.

19. See Firth, Hurst, and Hall, *Clinical Genetics*.

20. Alan E. Guttmacher, Jean Jenkins, and Wendy R. Uhlmann, "Genomic Medicine: Who Will Practice It? A Call to Open Arms," *American Journal of Medical Genetics (Seminars in Medical Genetics)* 106, no. 3 (2001): 216–22.

21. Dorothy Nelkin and M. Susan Lindee, *The DNA Mystique: The Gene as Cultural Icon*, 2nd ed. (Ann Arbor: University of Michigan Press, 2004).

22. Abby Lippman, "Prenatal Genetic Testing and Screening: Constructing Needs and Reinforcing Inequities," *American Journal of Law and Medicine* 17, no. 1/2 (1991): 15–50.

23. Judith Graham, "FDA Aims to Regulate Personal Genetic Tests," *Chicago Tribune*, June 11, 2010, www.chicagotribune.com/business/ct-biz-0612-brf1-genetic -tests-20100611,0,7350321.story.

24. Nicholas Wade, "A Decade Later, Genetic Map Yields Few Clues," *New York Times*, June 13, 2010, A1.

25. Keith Wailoo and Stephen Pemberton, *The Troubled Dream of Genetic Medicine: Ethnicity and Innovation in Tay-Sachs, Cystic Fibrosis, and Sickle Cell Disease* (Baltimore: Johns Hopkins University Press, 2006).

26. Catherine Wang, Richard Gonzalez, and Sofia D. Merajver, "Assessment of Genetic Testing and Related Counseling Services: Current Research and Future Directions," *Social Science & Medicine* 58, no. 7 (2004): 1427–42.

27. Erik Parens and Adrienne Asch, eds., *Prenatal Testing and Disability Rights* (Washington, DC: Georgetown University Press, 2000); Douglas C. Baynton, "Disability and the Justification of Inequality in American History," in *The New Disability History: American Perspectives*, ed. Paul K. Longmore and Lauri Umansky (New York: New York University Press, 2001), 33–57.

28. Maxwell J. Mehlmann, *The Price of Perfection: Individual and Society in the Era of Biomedical Enhancement* (Baltimore: Johns Hopkins University Press, 2009); Jürgen Habermas, *The Future of Human Nature* (Cambridge: Polity, 2003); Barbara Katz Rothman, *The Tentative Pregnancy: How Amniocentesis Changes the Experience of Motherhood* (New York: W. W. Norton, 1993); Joan Rothschild, *The Dream of the Perfect Child* (Bloomington: Indiana University Press, 2005).

29. Sarah Franklin and Celia Roberts, *Born and Made: An Ethnography of Preimplantation Genetic Diagnosis* (Princeton, NJ: Princeton University Press, 2006).

30. Regina H. Kenen and Ann C. M. Smith, "Genetic Counseling for the Next 25 Years: Models for the Future," *Journal of Genetic Counseling* 4, no. 2 (1995): 118.

31. Barbara Biesecker, interviewed by the author, February 1, 2007; Patricia McCarthy Beach, Dianne M. Bartels, and Bonnie S. LeRoy, "Commentary on Genetic Counseling—a Profession in Search of Itself," *Journal of Genetic Counseling* 11, no. 3 (2002): 187–91.

32. Robin Grubs, interviewed by the author, June 14, 2010.

33. Beverly Yashar, interviewed by the author, January 3, 2011; Kelly Ormond, "NSGC Foundations—Then, Now, and Tomorrow," *Journal of Genetic Counseling* 14, no. 2 (2005): 85–88; Steven Keiles, "2008 National Society of Genetic Counselors Presidential Address: The NSGC Should Do Something About That . . . and We Are," *Journal of Genetic Counseling* 18 (2009): 105–8; Wendy R. Uhlmann, "1999 Presidential Address to the National Society of Genetic Counselors," *Journal of Genetic Counseling* 9, no. 1 (2000): 3–8; Jessica L. Mester et al., "Perceptions of Licensure: A Survey of Michigan Genetic Counselors," *Journal of Genetic Counseling* 18 (2009): 357–65. For the most recent information on state licensure of genetic counselors see www.nsgc.org/Advocacy/StatesIssuingLicensesforGeneticCounselors/tabid/347/Default.aspx.

34. Robert G. Resta, "Defining and Redefining the Scope and Goals of Genetic Counseling," *American Journal of Medical Genetics Part C (Seminars in Medical Genetics)* 142C (2006): 269–75; Robin L. Bennett, "Leading Voices and the Power of One: 2002 Presidential Address to the National Society of Genetic Counselors," *Journal of Genetic Counseling* 12, no. 2 (2003): 97–107.

35. Robert Resta et al., "A New Definition of Genetic Counseling: National Society of Genetic Counselors' Task Force Report," *Journal of Genetic Counseling* 15, no. 2 (2006): 77–83.

36. Ibid.

37. Barbara B. Biesecker and Kathryn F. Peters, "Process Studies in Genetic Counseling: Peering into the Black Box," *American Journal of Medical Genetics (Seminars in Medical Genetics)* 106 (2001): 194.

38. Ibid.

39. Marni J. Falk et al., "Medical Geneticists' Duty to Warn At-Risk Relatives for Genetic Disease," *American Journal of Medical Genetics* 120A (2003): 374–80; R. Beth Dugan et al., "Duty to Warn At-Risk Relatives for Genetic Disease: Genetic Counselors' Clinical Experience," *American Journal of Medical Genetics* Part C 119C (2003): 27–34.

40. Report of an Ad Hoc Committee on Genetic Counseling to the American Society of Human Genetics, October 29, 1971, Correspondence, Charles J. Epstein folder, Curt Stern Papers (CS), MS Coll 5, American Philosophical Society Library (APSL), Philadelphia, PA.

41. Charles J. Epstein to Members, Committee on Genetic Counseling, February 3, 1972, Correspondence, Charles J. Epstein folder, CS, APSL. Notably, Margery Shaw, Margaret Thompson, and Marian Rivas accepted Epstein's invitation to join the Ad Hoc Committee, while Bearn, Graham, Stern, and McKusick appear to have cycled off the committee.

42. Memo from Charles J. Epstein to Members, Committee on Genetic Counseling, February 3, 1972, Correspondence, CS, Charles J. Epstein folder, APSL.

43. Charles J. Epstein, "A Position Paper on the Organization of Genetic Counseling," in *Genetic Counseling*, ed. Herbert A. Lubs and Felix de la Cruz (New York: Raven Press, 1977), 333–48.

44. Charles J. Epstein, interviewed by the author, May 15, 2008.

45. Kurt Hirschhorn, interviewed by the author, March 9, 2008; Audrey Heimler, "An Oral History of the National Society of Genetic Counselors," *Journal of Genetic Counseling* 6, no. 3 (1997): 315–36.

46. Marian Rivas, interviewed by the author, September 28, 2010.

47. Epstein, interview; Kurt Hirschhorn, interview; Jon Weil, interviewed by the author, November 20, 2007; Ann P. Walker and Diane Baker, "In Memoriam: Dr. Charles J. Epstein (1933–2011)," *Journal of Genetic Counseling* 20, no. 5 (2011): 425–28.

48. This definition was approved in 1974 and first included in F. C. Fraser, "Genetic Counseling," *American Journal of Human Genetics* 26 (1974): 636–59; and formally published by the Ad Hoc Committee on Genetic Counseling in "Genetic Counseling," *American Journal of Human Genetics* 27 (1975): 240.

49. Regina H. Kenen, "Genetic Counseling: The Development of a New Interdisciplinary Occupation Field," *Social Science & Medicine* 18, no. 7 (1984): 546.

50. "Genetic Counseling," *American Journal of Human Genetics* 27 (1975): 241.

51. Joan H. Marks and Melissa L. Richter, "The Genetic Associate: A New Health Professional," *American Journal of Public Health* 66, no. 4 (1976): 388–90.

52. Resta, "Defining and Redefining," 273.

53. Nathaniel Comfort, *The Science of Human Perfection: Heredity, Health, and Human Improvement in American Biomedicine* (New Haven, CT: Yale University Press, 2012).

54. Diane B. Paul, *The Politics of Heredity: Essays on Eugenics, Biomedicine, and the Nature-Nurture Debate* (Albany: State University of New York Press, 1998), 141.

55. See Daniel J. Kevles, *In the Name of Eugenics: Genetics and the Uses of Human Heredity*, 2nd ed. (Cambridge, MA: Harvard University Press, 1995).

56. Alexandra Minna Stern, *Eugenic Nation: Faults and Frontiers of Better Breeding in Modern America* (Berkeley: University of California Press, 2005).

57. Comfort, *Science of Human Perfection*.

58. Last Will and Testament, Charles Fremont Dight, Box 2, University of Minnesota Dight Institute of Human Genetics Records (DIHGR), University of Minnesota Archives (UMA), Minneapolis, Minnesota; also see Neal Ross Holtan, "From Eugenics to Public Health Genetics in Mid-Twentieth Century Minnesota" (PhD dissertation, University of Minnesota, 2011).

59. William Allan, "Notes on Negative Eugenics," Folder: Lectures Given by Allan, Box P323–1, William Allan Papers (WAP), Dorothy Carpenter Medical Archives (DCMA), Wake Forest Baptist Medical Center (WFBMC), Winston-Salem, North Carolina; and Nathaniel Comfort, "'Polyhybrid Heterogeneous Bastards': Promoting Medical Genetics in America in the 1930s and 1940s," *Journal of the History of Medicine and Allied Sciences* 61, no. 4 (2006): 415–55.

60. Robert G. Resta, "The Historical Perspective: Sheldon Reed and 50 Years of Genetic Counseling," *Journal of Genetic Counseling* 6, no. 4 (1997): 375–77.

61. Sheldon C. Reed, "The Genetic Counseling Explosion" (1972), Box 5, DIHGR, UMA. Note: after I consulted this collection in 2005, it was re-catalogued and reorganized into five boxes. My endnotes include document locations based on the detailed information in the new finding aid. Also see "Medicine: Sterilization and Heredity," *Time*, April 15, 1957, www.time.com/time/magazine/article/0,9171,862556,00.html.

62. S. C. Reed, "Genetic Counseling Explosion."

63. S. C. Reed, "The Delivery of Genetic Counseling," Box 4, DIHGR, UMA.

64. S. C. Reed, "Practical Genetic Counseling," Box 5, DIHGR, UMA: See Sheldon C. Reed, *Counseling in Medical Genetics* (Philadelphia: W. B. Saunders, 1955); also see Robert G. Resta, "In Memoriam: Sheldon Clark Reed, PhD, 1910–2003," *Journal of Genetic Counseling* 12, no. 3 (2003): 283–85.

65. See S. C. Reed to Harry L. Shapiro (President, American Eugenics Society), Box 2, DIHGR, UMA.

66. S. C. Reed, "Where Is Eugenics?," (1953), Box 5, DIHGR, UMA. Also see the speech "Human Betterment" (1979), which Reed delivered to the Human Betterment League of North Carolina in 1979, Box 5, DIHGR, UMA.

67. S. C. Reed, "The Significance of Genetic Counseling" (1967), Box 5, DIHGR, UMA.

68. S. C. Reed to Dr. Harry L. Shapiro, May 15, 1961, Box 3, DIHGR, UMA.

69. S. C. Reed, "Human Betterment."

70. Molly Ladd-Taylor notes Reed's revelation in her insightful analysis "'A Kind of Genetic Social Work': Sheldon Reed and the Origins of Genetic Counseling," in *Women, Health, and Nation: Canada and the United States since 1945*, ed. Georgina Feldberg et al. (Montreal: McGill University Press, 2003), 67–83.

71. S. C. Reed, "Genetic Counseling Explosion"; also see his "Genetic Counseling and Human Values" (1973), Box 5, DIHGR, UMA.

72. S. C. Reed, "Significance of Genetic Counseling."

73. S. C. Reed to Mr. Frederick Osborn, May 19, 1961, Box 2, DIHGR, UMA.

74. C. Nash Herndon, "Eugenics and Preventive Medicine," *Eugenical News* 38, no. 3 (September 1953): 56.

75. Frederick Osborn to James Neel, June 27, 1957, Osborn Correspondence, Box Oh-Pa, James V. Neel Papers (JVN), MS Coll 96, APSL.

76. Frederick Osborn to S. C. Reed, June 27, 1957, Box 2, DIHGR, UMA.

77. "Heredity Counseling Symposium," November 1, 1957, Box Oh-Pa, MS Coll 96, JVN, APSL.

78. Neel to Osborn, November 14, 1957, Box Oh-Pa, JVN, APSL.

79. Helen G. Hammons, ed., *Heredity Counseling* (New York: Hoeber-Harper, 1959).

80. Osborn to Neel, June 27, 1957, Box Oh-Pa, JVN, APSL.

81. Franz J. Kallmann, "Types of Advice Given by Heredity Counselors: I," chap. 9 in Hammons, *Heredity Counseling*, 82.

82. S. C. Reed, "Types of Advice Given to Heredity Counselors: II," chap. 10 in Hammons, *Heredity Counseling*, 92.

83. Ian H. Porter, "Evolution of Genetic Counseling in America," in *Genetic Counseling*, ed. Lubs and de la Cruz, 17–34.

84. M. Susan Lindee, *Moments of Truth in Genetic Medicine* (Baltimore: Johns Hopkins University Press, 2005); Comfort, "'Polyhybrid Heterogeneous Bastards.'"

85. Diane B. Paul, "The History of Newborn Phenylketonuria Screening in the U.S.," Appendix 5, in *Promoting Safe and Effective Genetic Testing in the United States: Final Report of the Task Force on Genetic Testing*, ed. Neil A. Holtzman and Michael S. Watson (Baltimore: Johns Hopkins University Press, 1999), 137–69; Paul, "Towards a Realistic Assessment of PKU Screening," *PSA: Proceedings of the Biennial Meeting of the Philosophy of Science Association* 2 (1994): 322–28.

86. Wendy Kline, *Bodies of Knowledge: Sexuality, Reproduction, and Women's Health in the Second Wave* (Chicago: University of Chicago Press, 2010).

87. Audrey Heimler, "An Oral History of the National Society of Genetic Counselors," *Journal of Genetic Counseling* 6, no. 3 (1997): 315–35.

88. Robert Resta, interviewed by the author, March 20, 2007; Robert Resta, "The Great Genetic Counseling Divorce of 1992: A Historical Perspective on Change in the Genetic Counseling Profession," *The DNA Exchange* (blog), April 11, 2010, http://thednaexchange.com.

89. Jennifer A. Bubb and Anne L. Matthews, "What's New in Prenatal Screening and Diagnosis," *Primary Care: Clinics in Office Practice* 31 (2004): 561–82.

90. James R. Sorenson and Arthur J. Culbert, "Genetic Counselors and Counseling Orientations—Unexamined Topics in Evaluation," in *Genetic Counseling*, ed. Lubs and de la Cruz, 131–56; "Who Should Counsel?—A Special Report," Reference-Articles-Genetic Counseling, 1969–1975, Human Genetics Graduate Program Records (HGGPR), Sarah Lawrence College (SLC), Bronxville, New York.

91. Joy Schaleben Lewis, "Genetic Counselors Multiply," *New York Times*, March 23, 1986, 22W/C–23W/C; Elizabeth M. Fowler, "Careers: Need Grows for Genetic Counselors," *New York Times*, June 5, 1990, 24; J. A. Scott et al., "Genetic Counselor Training: A Review and Considerations for the Future," *American Journal of Human Genetics* 42, no. 1 (1988): 191–99.

92. See Kenen and Smith, "Genetic Counseling for the Next 25 Years," 121.

93. Barton Childs, "Genetic Medicine," n.d., 14, Correspondence, Barton Childs folder, Curt Stern Papers (CS), MS Coll 5, APSL.

94. Ronald Conley and Aubrey Milunsky, "The Economics of Prenatal Genetic Diagnosis," chap. 20 in *The Prevention of Genetic Disease and Mental Retardation*, ed. Aubrey Milunsky (Philadelphia: W. B. Saunders, 1975), 448–49.

95. Available at www.nsgc.org.

96. Vivian Ota Wang, "Multicultural Genetic Counseling: Then, Now, and in the 21st Century," *American Journal of Medical Genetics (Seminars in Medical Genetics)* 106, no. 3 (2001): 208–15; Ilana Suez Mittman and Katy Downs, "Diversity in Genetic Counseling: Past, Present and Future," *Journal of Genetic Counseling* 17, no. 4 (2008): 301–13; Stephanie C. Smith, Nancy Steinberg Warren, and Lavanya Misra, "Minority Recruitment into the Genetic Counseling Profession," *Journal of Genetic Counseling* 2, no. 3 (1993): 171–81; Tracey Oh and Linwood J. Lewis, "Consideration of Genetic Counseling as a Career: Implications for Diversifying the Genetic Counseling Field," *Journal of Genetic Counseling* 14, no. 1 (2005): 71–81.

97. Nancy Callanan, "2005 National Society of Genetic Counselors Presidential Address: Raising Our Voice," *Journal of Genetic Counseling* 15, no. 2 (2006): 74.

98. Jon Weil, interviewed by the author, November 23, 2007.

99. Resta, interview.

100. Kathryn T. Hock et al., "Direct-to-Consumer Genetic Testing: An Assessment of Genetic Counselors' Knowledge and Beliefs," *Genetics in Medicine* 13, no. 4 (2011): 325–32.

101. United States Government Accountability Office, *Direct-to-Consumer Genetic Tests: Misleading Test Results Are Further Complicated by Deceptive Marketing and Other Questionable Practices*. GAO-10-847, July 22, 2010, www.gao.gov/products/GAO-10-847T; Shobita Parthasarathy, *Building Genetic Medicine: Breast Cancer, Technology, and the Comparative Politics of Health Care* (Cambridge, MA: MIT Press, 2007); Cheryl Berg and Kelly Fryer-Edwards, "The Ethical Challenges of Direct-to-Consumer Genetic Testing," *Journal of Business Ethics* 77, no. 1 (2007): 17–31.

CHAPTER TWO: *Genetic Risk*

1. "Huntington Disease (HD)," in Helen V. Firth, Jane A. Hurst, and Judith G. Hall, *Oxford Desk Reference: Clinical Genetics* (Oxford: Oxford University Press, 2005), 354–56; Alice Wexler, *The Woman Who Walked into the Sea: Huntington's and the Making of a Genetic Disease* (New Haven, CT: Yale University Press, 2008).

2. Kindred 261, University of Michigan Adult Medical Genetics Clinic Records (AMGCR), Collection 4446 Bimu, Bentley Historical Library (BHL), University of Michigan (UM). Access approved by University of Michigan Institutional Review Board, HUM00012519. Only pseudonyms are used to protect patient confidentiality.

3. This kindred first came to the attention of Estella M. Hughes of Smith College when she was researching her 1923 social work thesis at Smith College. After graduation, Hughes became a psychiatric social worker at Kalamazoo State Hospital, and later she was hired as part of the Heredity Clinic's initial staff. See Estella M. Hughes, "Social Significance of Huntington's Chorea," *American Journal of Psychiatry* 4, no. 3 (1925): 537–74.

4. Note from August 1985, Kindred 261, AMGCR, BHL, UM.

5. Department of Health, Education, and Welfare, Public Health Service, National Institutes of Health, *Report: Commission for the Control of Huntington's Disease and Its Consequences*, Vol. 1, Overview, October 1977, 2.

6. Nancy Sabin Wexler, "Genetic 'Russian Roulette': The Experience of Being 'At Risk' for Huntington's Disease," chap. 12 in *Genetic Counseling: Psychological Dimensions*, ed. Seymour Kessler (New York: Academic Press, 1979), 218.

7. Alice Wexler, *Mapping Fate: A Memoir of Family, Risk, and Genetic Research* (Berkeley: University of California Press, 1995).

8. James F. Gusella et al., "A Polymorphic DNA Marker Genetically Linked to Huntington's Disease," *Nature* 306 (November 17, 1983): 234–38.

9. Susan E. Andrew et al., "The Relationship between Trinucleotide (CAG) Repeat Length and Clinical Features of Huntington's Disease," *Nature Genetics* 4 (August 1993): 398–403; The Huntington's Disease Collaborative Research Group, "A Novel Gene Containing a Trinucleotide Repeat That Is Expanded and Unstable on Huntington's Disease Chromosomes," *Cell* 72 (March 26, 1993): 971–83.

10. Department of Health, Education, and Welfare, Public Health Service, Na-

tional Institutes of Health, *Report: Commission for the Control of Huntington's Disease*.

11. Robert Klitzman et al., "Decision-Making about Reproductive Choices among Individuals At-Risk for Huntington's Disease," *Journal of Genetic Counseling* 16, no. 3 (2007): 347–62.

12. David H. Smith et al., *Early Warning: Cases and Ethical Guidance for Pre-symptomatic Testing in Genetic Diseases* (Bloomington and Indianapolis: Indiana University Press, 1998).

13. Wexler, *Mapping Fate*, 234.

14. Smith et al., *Early Warning*.

15. Ian D. Young, *Introduction to Risk Calculation in Genetic Counseling*, 3rd ed. (Oxford: Oxford University Press, 2007).

16. Jennifer Roggenbuck et al., "Perception of Genetic Risk among Genetic Counselors," *Journal of Genetic Counseling* 9, no. 1 (2000): 47–59.

17. Bonnie J. Rough, *Carrier: Untangling the Danger in My DNA* (Berkeley: Counterpoint, 2010).

18. Nicole Hahn Rafter, *White Trash: The Eugenic Family Studies, 1877–1919* (Boston: Northeastern University Press, 1988); Alexandra Minna Stern, "'We Cannot Make a Silk Purse out of a Sow's Ear': Eugenics in the Hoosier Heartland," *Indiana Magazine of History* 103, no. 1 (2007): 3–38.

19. "The Early Life of Lee Raymond Dice," Biographical Binder, Lee Raymond Dice Papers (LRD), Collection Aa2, BHL, UM.

20. "Professional History of Lee Raymond Dice," Biographical Binder, LRD, BHL, UM.

21. Ibid.

22. "History of Department of Human Heredity," "Report to Dean Yoakum on Research Grant R-108 (L. R. Dice), 1940–1941," and "The Heredity Clinic, September 20, 1941," Box 1, Department of Human Genetics Records (DHGR), BHL, UM.

23. Minutes of the Staff Meeting, Laboratory of Vertebrate Biology, June 24, 1941, Box 4, LRD, BHL, UM.

24. "History of Department of Human Heredity," and Lee R. Dice to Dr. Richard J. Porter, January 10, 1956, Box 1, DHGR, BHL, UM.

25. Clinic materials in Box 2, AMGCR, BHL, UM.

26. Office Procedures Manual, Box 1, DHGR, BHL, quotes from "Reports" section.

27. Robert G. Resta, "The Crane's Foot: The Rise of the Pedigree in Human Genetics," *Journal of Genetic Counseling* 2, no. 4 (1993): 256; Robert G. Resta, "Genetic Drift: Whispered Hints," *American Journal of Medical Genetics* 59 (1995): 131–33.

28. Lee R. Dice, "Symbols for Human Pedigree Charts," *Journal of Heredity* 37, no. 1 (January 1946): 11–15.

29. James V. Neel, *Physician to the Gene Pool: Genetic Lessons and Other Stories* (New York: John Wiley and Sons, 1994), 17.

30. Sheldon C. Reed, "Human Factors in Genetic Counseling," reprint, source unspecified, 107, Box 5, DIHGR, UMA.

31. S. C. Reed, Untitled talk (begins "Thank you, Dr. Jensen," n.d., likely 1960s), likely in Box 4, DIHGR, UMA.

32. S. C. Reed, Selected Counseling Cases, Box 2; and "Incest Children: Brother x Sister Unions," Box 3, DIHGR, UMA.

33. S. C. Reed, Public Services of the Dight Institute, n.d. but soon after his arrival, Box 2, DIHGR, UMA.

34. S. C. Reed, "Etiology and Genetic Counseling in Mental Retardation" (1955), Box 4, DIHGR, UMA.

35. See Kindreds 95 and 106, AMGCR, BHL, UM.

36. S. C. Reed, "The Significance of Genetic Counseling" (1967), 11, Box 5, DIHGR, UMA.

37. Selected Counseling Cases, n.d. but likely 1960s, Box 2, DIHGR, UMA.

38. Drawn from analysis of first 400 Case Record Sheets, Box 64, AMGCR, BHL, UM.

39. Robert Albrook, "U. Geneticists Seek Aid in Fight on Breast Cancer," *Minneapolis Morning Tribune*, January 9, 1946, n.p.; "Expert Finds Cancer Clue: Sisters of Patients Held Susceptible," *Minneapolis Sunday Tribune*, April 16, 1950, Box 2, DIHGR, UM.

40. Deborah Lupton, *Risk* (London: Routledge, 1999), 5.

41. See Ian Hacking, *The Taming of Chance* (Cambridge: Cambridge University Press, 1990).

42. Ulrich Beck, *Risk Society: Towards a New Modernity* (London: Sage, 1992), 21; also see William G. Rothstein, *Public Health and the Risk Factor: A History of An Uneven Medical Revolution* (Rochester, NY: University of Rochester Press, 2003).

43. Paul Slovic, *The Perception of Risk* (Sterling, VA: Earthscan, 2000); Baruch Fischoff et al., *Acceptable Risk* (Cambridge: Cambridge University Press, 1981).

44. James E. Short Jr., "The Social Fabric at Risk: Toward the Social Transformation of Risk Analysis," *American Sociological Review* 49, no. 6 (1984): 711–25.

45. Chauncey Starr, Richard Rudman, and Chris Whipple, "Philosophical Basis for Risk Analysis," *Annual Review in Energy* 1 (1976): 629–62; Chris Whipple and Vincent T. Covello, eds., *Risk Analysis in the Private Sector* (New York: Plenum Press, 1985).

46. Chauncey Starr and Chris Whipple, "A Perspective on Health and Safety Risk Analysis," *Management Science* 30, no. 4 (1984): 452–63; Chauncey Starr, "Risk Management, Assessment, and Acceptability," *Risk Analysis* 5, no. 2 (1985): 97–102; Elke U. Weber, "A Descriptive Measure of Risk," *Acta Psychologica* 69 (1988): 185–203; Stanley Kaplan and B. John Garrick, "On the Quantitative Definition of Risk," *Risk Analysis* 1, no. 1 (1981): 11–27.

47. S. C. Reed, "Etiology and Genetic Counseling."

48. Christina G. S. Palmer and François Sainfort, "Toward a New Conceptualization and Operationalization of Risk Perception within the Genetic Counseling Domain," *Journal of Genetic Counseling* 2, no. 4 (1993): 278.

49. Charles Vlek, "Risk Assessment, Risk Perception and Decision Making about Courses of Action Involving Genetic Risk: An Overview of Concepts and Methods," *Birth Defects: Original Article Series* 23, no. 2 (1987): 180; Patrick Humphreys and Dina Berkeley, "Representing Risks: Supporting Genetic Counseling," *Birth Defects: Original Article Series* 23, no. 2 (1987): 227–50.

50. J. H. Pearn, "Patients' Subjective Interpretation of Risks Offered in Genetic Counselling," *Journal of Medical Genetics* 10, no. 2 (1973): 131.

51. F. Clarke Fraser, "Counseling in Genetics: Its Intent and Scope," *Birth Defects: Original Article Series* 6, no. 1 (1970): 10.

52. Abby Lippman-Hand and F. Clarke Fraser, "Genetic Counseling: Provision and Reception of Information," *American Journal of Medical Genetics* 3 (1979): 113–27; A. Lippman-Hand and F. Clarke Fraser, "Genetic Counseling—the Postcounseling Period: I. Parents' Perceptions of Uncertainty," *American Journal of Medical Genetics* 4 (1979): 51–71; A. Lippman-Hand and F. Clarke Fraser, "Genetic Counseling—the Postcounseling Period: II. Making Reproductive Choices," *American Journal of Medical Genetics* 4 (1979): 73–87.

53. Abby Lippman-Hand and F. Clarke Fraser, "Genetic Counseling: Parents' Responses to Uncertainty," *Birth Defects: Original Article Series* 15, no. 5C (1979): 331.

54. Jon Weil, *Psychosocial Genetic Counseling* (Oxford: Oxford University Press, 2000), 133–34.

55. Lippman-Hand and Fraser, "Genetic Counseling: Provision and Reception," 124.

56. Christina G. S. Palmer and François Sainfort, "Toward a New Conceptualiztion and Operationalization of Risk Perceptive within the Genetic Counseling Domain," *Journal of Genetic Counseling* 2, no. 4 (1993): 288.

57. Sheila Jasanoff, "Bridging the Two Cultures of Risk Analysis," *Risk Analysis* 13, no. 2 (1993): 123–29.

58. Paul Slovic, ed., *The Feeling of Risk: New Perspectives on Risk Perception* (London: Earthscan, 2010).

59. Carlos Novas and Nikolas Rose, "Genetic Risk and the Birth of the Somatic Individual," *Economy and Society* 29, no. 4 (2000): 485–513; Martha Lampland and Susan Leigh Star, eds., *Standards and Their Stories: How Quantifying, Classifying, and Formalizing Practices Shape Everyday Life* (Ithaca, NY: Cornell University Press, 2009).

60. Mary Douglas and Aaron Wildavsky, *Risk and Culture: An Essay on the Selection of Technological and Environmental Dangers* (Berkeley: University of California Press, 1982).

61. Barbara Katz Rothman, *The Tentative Pregnancy: How Amniocentesis Changes the Experience of Motherhood*, 2nd ed. (New York: W. W. Norton, 1993), 43.

62. Diana Puñales-Morejon, interviewed by the author, January 19, 2007.

63. S. C. Reed, "The Delivery of Genetic Counseling," Box 4, DIHGR, UMA.

64. L. S. Penrose, "The Relative Effects of Paternal and Maternal Age in Mongolism," *Journal of Genetics* 27 (1933): 219–24.

65. J. A. Böök and S. C. Reed, "Empiric Risk Figures in Mongolism," *Journal of the American Medical Association* 143, no. 8 (June 24, 1950): 730–32.

66. Ibid., 732.

67. See Hans Olof Åkesson, "Empiric Risk Figures in Mental Deficiency," *Acta Genetica* 12 (1962): 28–32; J. G. Masterson, "Empiric Risk, Genetic Counseling and Preventive Measures in Anencephaly," *Acta Genetica* 12 (1962): 219–29.

68. Robert Albrook, "U. Geneticists Seek Aid in Fight on Breast Cancer," *Minneapolis Morning Tribune*, January 9, 1946, n.p.; "Expert Finds Cancer Clue."

69. S. C. Reed to Dr. Charles O. Warren, November 22, 1948, and "'U' Cancer Experts Get $55,352 Grants," *Minneapolis Morning Tribune*, April 20, 1949, n.d., Box 2, DIHGR, UMA.

70. "With the American Cancer Society Research Tour, Minneapolis, Humans Do Not Inherit Cancer," undated press release, Box 2, DIHGR, UMA.

71. "Expert Finds Cancer Clue," 2.

72. Böök and Reed, "Empiric Risk Figures in Mongolism," 732.

73. James V. Neel, "The Meaning of Empiric Risk Figures for Disease or Defect," *Eugenics Quarterly* 5, no. 1 (March 1958): 41–42.

74. Kenneth K. Kidd, "Empiric Recurrence: Risks and Models of Inheritance: Part II," *Birth Defects: Original Article Series* 15, no. 5C (1979): 51; Nancy Role Mendell and M. Anne Spence, "Empiric Recurrence: Risks and Models of Inheritance: Part I," *Birth Defects: Original Article Series* 15, no. 5C (1979): 39–49; Seymour Packman, "Empiric Risk Counseling: A Perspective," *Birth Defects: Original Article Series* 15, no. 5C (1979): 69–70.

75. Jon Weil, e-mail message to the author, May 20, 2011.

76. André Rogatko, "Evaluating the Uncertainty of Risk Prediction in Genetic Counseling: A Bayesian Approach," *American Journal of Human Genetics* 31 (1988): 513–19.

77. S. C. Reed, "Practical Genetic Counseling," Box 5, DIHGR, UMA.

78. Kathleen and Arthur S. Postle, "Whose Little Girl Are You?," *McCall's*, December 1948, 22–23, 126, 134, 138.

79. Minutes, Heredity Clinic, December 3, 1948, Folder: Laboratory of Vertebrate Biology, Meeting Minutes, 1945–1952, LRD, BHL, UM.

80. Robert J. R. Johnson, "Heredity Explained—U Team Calls Toss on Genetic Gamble," *Pioneer Press*, March 16, 1958, n.p., Box 2, DIHGR, UMA.

81. "Mr. Fixit Says: No Wisecrack, Blue Eyes," *Minneapolis Morning Tribune*, November 17, 1948, 17; "Mr. Fixit Says: Cousins Can't Wed Here," *Minneapolis Morning Tribune*, May 19, 1949, 19; "How to Join Baldy Row," *Minneapolis Morning Tribune*, May 6, 1953, 12, Box 2, DIHGR, UMA.

82. Sheldon C. Reed, *Counseling in Medical Genetics* (Philadelphia: W. B. Saunders, 1955).

83. Ibid., 12.

84. "Got Genes? Then You Must Read This Book," *Minneapolis Sunday Tribune*, October 2, 1955, 2, Box 3, DIHGR, UMA.

85. Specimen advertising page from W. B. Saunders Company, Box 3, DIHGR, UMA.

86. S. C. Reed, "Genetic Counseling (Women's Studies)" (1977), likely Box 5, DIHGR, UMA.

87. Frances Whitney to S. C. Reed, January 14, 1957, Box 2, DIHGR, UMA.

88. Alice Wexler, "Stigma, History, and Huntington's Disease," *Lancet* 376 (June 30, 2010): 18–19, doi:10.1016/S0140-6736(10)60957-9.

89. Mrs. Nedra C. Kuntz to X (name withheld), August 8, 1944, Kindred 261, AMGCR, BHL, UM.

90. Mrs. L to Ms. Hughes, December 19, 1955, Kindred 302, AMGCR, BHL, UM.

91. S. C. Reed, Untitled talk at Dight Institute, probably 1960s, Box 3, DIHGR, UMA.

92. S. C. Reed, Untitled talk ("Thank you, Dr. Jensen").

93. Ian D. Young, *Introduction to Risk Calculation in Genetic Counseling*, 3rd ed. (Oxford: Oxford University Press, 2007).

94. S. C. Reed, Untitled talk ("Thank you, Dr. Jensen").

95. S. C. Reed, "Practical Genetic Counseling."

96. S. C. Reed to James R. Sorenson, November 2, 1970, Box 2, DIHGR, UMA.

97. Richard D. James, "Genetic Counselors Give Advice to Parents-to-be on Hereditary Defects," *Wall Street Journal*, October 5, 1967, n.p., Box 2, DIHGR, UMA.

98. William Leeming, "Tracing the Shifting Sands of 'Medical Genetics': What's in a Name?," *Studies in History and Philosophy of Biological and Biomedical Sciences* 41 (2010): 50–60.

99. S. C. Reed, "Medical and Genetic Indications for Therapeutic Abortion," Box 3, DIHGR, UMA.

100. Roggenbuck et al., "Perception of Genetic Risk."

101. Dorothy C. Wertz, James R. Sorenson, and Timothy C. Heeren, "Clients' Interpretation of Risks Provided in Genetic Counseling," *American Journal of Human Genetics* 39 (1986): 253–64.

102. Katie Featherstone et al., *Risky Relations: Family, Kinship, and the New Genetics* (Oxford: Berg, 2006), 16.

103. "Huntington Disease (HD)," in Firth, Hurst, and Hall, *Clinical Genetics*, 354–56; Wexler, *Woman Who Walked into the Sea*.

104. Note from August 1985, Kindred 261, AMGCR, BHL, UM.

105. See Patricia T. Kelly, *Dealing with Dilemma: A Manual for Genetic Counselors* (New York: Springer, 1977); Joan H. Marks et al., *Genetic Counseling Principles in Action: A Casebook* (White Plains, NY: March of Dimes Birth Defects Foundation, 1989); Jon Weil, *Psychosocial Genetic Counseling*.

106. Wexler, *Woman Who Walked into the Sea*; Beth Linnen, "Marjorie Guthrie Sings Out against Cause of Woody's Death," *Minnesota Daily* 78, no. 45 (October 14, 1977), n.p., Box 2, DIHGR, UM.

107. Marjorie Guthrie, "A Personal View of Genetic Counseling," in *Counseling in Genetics*, ed. Y. Edward Hsia et al. (New York: Alan R. Liss, 1979), 329–41.

108. Diane Baker, interviewed by the author, January 16, 2008.

109. Dorene Markel, interviewed by the author, December 7, 2007; similar efforts were underway at the University of Minnesota. See Joe Kimball, "With Help, They Face Huntington's Disease," Family Living Section, *Minneapolis Tribune*, Sunday, Feburary 10, 1980, 12F, Box 2, DIHGR, UMA.

110. Dorene Markel, e-mail message to the author, August 11, 2011.

111. Shelly Clark et al., "Patient Motivation, Satisfaction, and Coping in Genetic Counseling and Testing for BRCA1 and BRCA2," *Journal of Genetic Counseling* 9, no. 3 (2000): 219–35.

1. Interracial Service, Sunday, February 11, 1951, The Cathedral Church of St. Mark, Box 4, University of Minnesota Dight Institute for Human Genetics Records (DIHGR), University of Minnesota Archives (UMA), Minneapolis, Minnesota; Book 2, January 1, 1950–June 30, 1951, Dight Institute Inquiries, private collection held by the author, Ann Arbor, Michigan.

2. Sheldon Reed, "All Men Are Brothers under the Skin," delivered February 11, 1951, Box 4, DIHGR, UMA.

3. Book 2, January 1, 1950–June 30, 1951, Dight Institute Inquiries.

4. S. C. Reed, "A Scientist Looks at the Races of Man," Box 2, DIHGR, UMA.

5. S. C. Reed, "Color of the U.S.A.—3000 A.D.," (1954), Box 4, DIHGR, UMA.

6. Reed and Dobzhansky carried on a lively correspondence in the 1940s and 1950s; see letters between the two in Box 1, DIHGR, UMA.

7. Geoffrey C. Bowker and Susan Leigh Star, *Sorting Things Out: Classification and Its Consequences* (Cambridge, MA: MIT Press, 1999).

8. See Matthew Connelly, *Fatal Mis-Conception: The Struggle to Control World Population* (Cambridge, MA: Belknap Press of Harvard University Press, 2008).

9. See Dorothy Roberts, *Fatal Invention: How Science, Politics, and Big Business Re-Create Race in the Twenty-First Century* (New York: New Press, 2011); also see Pilar Ossorio, "Societal and Ethical Issues in Pharmacogenics," chap. 12 in *Pharmacogenomics: Applications to Patient Care* (Kansas City, MO: American College of Clinical Pharmacy, 2004), 339–439.

10. For scientific, sociological, and anthropoligical analyses of race and biology see Joseph L. Graves Jr., *The Emperor's New Clothes: Biological Theories of Race at the Millennium* (New Brunswick, NJ: Rutgers University Press, 2002); Troy Duster, "Buried Alive: The Concept of Race in Science," chap. 13 in *Genetic Nature/Culture: Anthropology and Science beyond the Two-Culture Divide*, ed. Alan H. Goodman, Deborah Heath, and M. Susan Lindee (Berkeley: University of California Press, 2003), 258–77; Anne Fausto-Sterling, "Refashioning Race: DNA and the Politics of Health Care," *differences: A Journal of Feminist Cultural Studies* 15, no. 3 (2004): 1–37; Jonathan Kahn, "Genes, Race, and Population: Avoiding a Collision of Categories," *American Journal of Public Health* 96, no. 11 (2006): 1965–70; Pilar N. Ossorio and Troy Duster, "Race and Genetics: Controversies in Biomedical, Behavioral, and Forensic Sciences," *American Psychologist* 60, no. 1 (2005): 115–28; Duana Fullwiley, "The Biologistical Construction of Race: 'Admixture' Technology and the New Genetic Medicine," *Social Studies of Science* 38, no. 5 (2008): 695–735.

11. See Daniel J. Kevles, *In the Name of Eugenics: Genetics and the Uses of Human Heredity*, 2nd ed. (Cambridge, MA: Harvard University Press, 1995); and Diane B. Paul, *Controlling Human Heredity: 1865 to the Present* (Amherst, NY: Humanity Books, 1995).

12. See Peggy Pascoe, *What Comes Naturally: Miscegenation Law and the Making of Race in America* (New York: Oxford University Press, 2009).

13. See Elazar Barkan, *The Retreat of Scientific Racism: Changing Concepts of Race in Britain and the United States between the World Wars* (New York: Cambridge

University Press, 1992); Nancy Stepan, *The Idea of Race in Science: Great Britain 1800-1960* (Hamden, CT: Archon Books, 1982); William B. Provine, "Geneticists and Race," *American Zoologist* 26 (1986): 857-87; Joanne Meyerowitz, "How Common Culture Shapes the Separate Lives: Sexuality, Race, and Mid-Twentieth-Century Social Constructionist Thought," *Journal of American History* 96, no. 4 (2010): 1057-84.

14. See Jenny Reardon, *Race to the Finish: Identity and Governance in an Age of Genomics* (Princeton, NJ: Princeton University Press, 2005); and Rachel Caspari, "From Types to Populations: A Century of Race, Physical Anthropology, and the American Anthropological Association," *American Anthropologist* 105, no. 1 (2003): 65-76.

15. Reardon, *Race to the Finish*, 37.

16. UNESCO, *The Race Concept: Results of an Inquiry* (Paris, 1952), 99. Sheldon Reed was one of several dozen experts invited to comment on the UNESCO race statements.

17. Also see Melinda Gormley, "Scientific Discrimination and the Activist Scientist: L. C. Dunn and the Professionalization of Genetics and Human Genetics in the United States," *Journal of the History of Biology* 42 (2009): 33-72.

18. L. C. Dunn and Th. Dobzhansky, *Heredity, Race, and Society*, rev. ed. (New York: New American Library, 1952), 115.

19. Ibid., 118, 126.

20. See Lisa Gannett, "Racism and Human Genome Diversity Research: The Ethical Limits of 'Population Thinking,'" *Philosophy of Science* 68, no. 3 (2001): S479-92.

21. Ibid., S490.

22. Reardon, *Race to the Finish*, 2.

23. See Roberts, *Fatal Invention*; Dorothy Nelkin and M. Susan Lindee, *The DNA Mystique: The Gene as Cultural Icon*, 2nd ed. (Ann Arbor, MI: University of Michigan Press, 2004); Troy Duster, *Backdoor to Eugenics*, 2nd ed. (New York: Routledge, 2003).

24. S. C. Reed, "Population Brakes" and "Dynamics of Population Control," Box 4; "Hunger—Winner in the Population Race?," *Minneapolis Star*, April 4, 1951, Box 2, DIHGR, UMA.

25. "Population Control Called Possible without Offending Religion," *Minneapolis Morning Tribune*, January 13, 1960, 13, Box 2, DIHGR, UMA.

26. S. C. Reed, "Toward a New Eugenics: The Importance of Differential Reproduction," *Eugenics Review* 57, no. 2 (1965): 74, Box 5, DIHGR, UMA.

27. One of the best studies of the ERO remains Garland E. Allen, "The Eugenics Record Office at Cold Spring Harbor, 1910-1940: An Essay in Institutional History," *Osiris*, 2nd series, 2 (1986): 225-64.

28. See Paul A. Lombardo, "'The American Breed': Nazi Eugenics and the Origins of the Pioneer Fund," *Albany Law Review* 65, no. 3 (2002): 743-830.

29. See William H. Tucker, *The Funding of Scientific Racism: Wickliffe Draper and the Pioneer Fund* (Urbana: University of Illinois Press, 2002); and Michael G. Kenney, "Toward a Racial Abyss: Eugenics, Wickliffe Draper, and the Origins of the

Pioneer Fund," *Journal of History of the Behavioral Sciences* 38, no. 3 (2002): 259–83.

30. James V. Neel to C. Nash Herndon, March 7, 1950, Box Correspondence He-Ho, James V. Neel Papers (JVN), MS Coll 96, American Philosophical Society Library (APSL), Philadelphia, Pennsylvania.

31. See his discussion of his attitudes toward eugenics in James V. Neel, *Physician to the Gene Pool: Genetic Lessons and Other Stories* (New York: John Wiley and Sons, 1994).

32. James V. Neel to Frederick Osborn, December 10, 1953, Box 4, American Eugenics Society (AES) folder, JVN, APSL.

33. Nicole Hahn Rafter, ed., *White Trash: The Eugenic Family Studies, 1877–1919* (Boston: Northeastern University Press, 1988).

34. Regents Communication, Exhibits of the Regents Meetings, May 20, 1950, Box 63, University of Michigan Board of Regents records (UMBR), Bentley Historical Library (BHL), University of Michigan (UM).

35. James P. Adams to Mr. Wickliffe Draper, April 25, 1950, Exhibits of the Regents Meetings, May 20, 1950, Box 63, UMBR, BHL, UM.

36. Wickliffe Draper to the President of the University of Michigan, April 18, 1950, Exhibits of the Regents Meetings, May 20, 1950, Box 63, UMBR, BHL, UM.

37. Institute of Human Biology, University of Michigan, Minutes of staff meeting of October 23, 1950, Folder 1949–1950, Box 2, University of Michigan Department of Human Genetics Records (DHGR), BHL, UM; J. N. Spuhler, "Assortative Mating with Respect to Physical Characteristics," *Eugenics Quarterly* 15, no. 2 (1968): 128–40.

38. James V. Neel to Dr. Sheldon Reed, March 28, 1950, Box 2, DIHGR, UMA.

39. See Kevin Begos, "Benefactor with a Racist Bent: Wealthy Recluse Apparently Liked the Looks and Potential of Bowman Gray's New Medical-Genetics Department," from Part Two of "Against Their Will: North Carolina's Sterilization Program" (2002), Web-based supplement, *Winston-Salem Journal*, http://extras.journalnow.com/againsttheirwill/parts/two/printstory3.html.

40. Nash Herndon to L. H. Snyder, March 10, 1950, Snyder Folder, Unprocessed Box, Nash Herndon and William Allan Papers (NHWAP), Dorothy Carpenter Medical Archives (DCMA), Wake Forest Baptist Medical Center (WFBMC), Winston-Salem, North Carolina. I thank Monica Garnett for sending me these and additional materials that she found after my research trip to Winston-Salem.

41. On Allan see Nathaniel Comfort, "'Polyhybrid Heterogeneous Bastards': Promoting Medical Genetics in America in the 1930s and 1940s," *Journal of the History of Medicine and Allied Sciences* 61, no. 4 (2006): 415–55; also see William Allan, "Medicine's Need of Eugenics," *Southern Medicine & Surgery* 98, no. 8 (1936): reprint with no page numbers, Folder: Journal Reprints, Box P323-1, Dr. William Allan Papers (WAP), DCMA, WFBMC.

42. Nash Herndon to Mrs. Allan, November 18, 1957, Allan Correspondence—Misc, Nash Herndon Papers (NH), DCMA, WFBMC.

43. "Suggested Program of Research in Medical Genetics," and "Review of the Year's Activity, July 1941–July 1942," Folder: Foundation-Carnegie, Unprocessed Box, Department of Medical Genetics Records (DMGR), DCMA, WFBMC.

44. "Suggested Program of Research in Medical Genetics," Unprocessed Box, Nash Herndon and William Allan Papers (NHWAP), DCMA, WFBMC.

45. "Review of the Year's Activity, July 1941–July 1942."

46. William Allan to Dr. Laurence H. Snyder, October 19, 1941, Unprocessed Box, NHWAP, DCMA, WFBMC.

47. Johanna Schoen, *Choice & Coercion, Birth Control, Sterilization, and Abortion in Public Health and Welfare* (Chapel Hill: University of North Carolina Press, 2005).

48. "Report of Department of Medical Genetics," May 1, 1943–May 1, 1944; also see "Report of Department of Medical Genetics," May 1, 1946–May 1, 1947, Folder: Foundation-Carnegie, Unprocessed Box, DMGR, DCMA, WFBMC.

49. Quoted in Begos, "Benefactor with a Racist Bent."

50. S. C. Reed to Mr. Ronald W. May, February 3, 1960; S. C. Reed to Mr. Ronald W. May, February 11, 1960, Box 2, DIHGR, UMA.

51. Alexandra Minna Stern, *Eugenic Nation: Faults and Frontiers of Better Breeding in Modern America* (Berkeley: University of California Press, 2005).

52. S. C. Reed to President James L. Morrill; R. Joel Tierney to Mr. Al Cheese, September 16, 1976; S. C. Reed to Dean Richard S. Caldecott, September 23, 1976, Box 2, DIHGR, UMA. There are several thick folders in the DIHGR that reveal the extensive correspondence between Goethe and Reed from the 1940s to the 1960s.

53. Ellen Herman, *Kinship by Design: A History of Adoption in the Modern United States* (Chicago: University of Chicago Press, 2008); Rickie Solinger, *Beggars and Choosers: How the Politics of Choice Shapes Adoption, Abortion, and Welfare in the United States* (New York: Hill and Wang, 2002); Rickie Solinger, *Wake Up Little Susie: Single Pregnancy and Race before* Roe v. Wade, 2nd ed. (New York: Routledge, 2000).

54. E. Wayne Carp, ed., *Adoption in America: Historical Perspectives* (Ann Arbor: University of Michigan Press, 2002).

55. Judith Modell and Naomi Dambacher, "Making a 'Real' Family: Matching and Cultural Biologism in American Adoption," *Adoption Quarterly* 1, no. 2 (1997): 16; also see Ellen Herman, "The Difference Difference Makes: Justine Wise Polier and Religious Matching in Twentieth-Century Child Adoption," *Religion and American Culture* 10, no. 1 (Winter 2000): 57–98.

56. Sheldon C. Reed, *Counseling in Medical Genetics* (Philadelphia: W. B. Saunders, 1955), 8.

57. Book 1, 1948–1949, Dight Institute Inquiries.

58. Ibid.

59. Consultations requested by Adoption Agencies, Dight Institute for Human Genetics, September 1947–December 1957, Box 3, DIHGR, UMA.

60. Reed, *Counseling in Medical Genetics*, 155.

61. Ibid., chap. 20; and S. C. Reed, "Human Racial Differences," Box 5, DIHGR, UMA.

62. C. Stern, "Model Estimates of the Number of Gene Pairs Involved in Pigmentation Variability of the Negro-American," *Human Heredity* 20 (1970): 165–68; Reed, "Human Racial Differences."

63. Book 2, January 1, 1950–June 30, 1951, Dight Institute Inquiries.

64. "Whites Have White Babies," *Minneapolis Morning Tribune*, August 20, 1949, 11, Box 2, DIHGR, UMA.

65. Adoption Referrals, Dight Institute, Box 3, DIHGR, UMA. Reed often addressed this issue in the popular media; see "Mr. Fixit Says: Whites Have White Babies," *Minneapolis Morning Tribune*, August 20, 1949, 11, Box 2, DIHGR, UMA.

66. Reed, "Adoption and the Child's Heritage" (1957), Box 4, DIHGR, UMA.

67. Reed, *Counseling in Medical Genetics*, 159; Sheldon C. Reed and Esther B. Nordlie, "Genetic Counseling for Children of Mixed Racial Ancestry," *Eugenics Quarterly* 8, no. 3 (1961): 157–63.

68. Reed, *Counseling in Medical Genetics*, 159.

69. Ibid., 158; also see "Illegitimate Child Held Good Risk for Adoption," *Minneapolis Morning Tribune*, November 4, 1949, n.p., Box 2, DIHGR, UMA.

70. Reed, *Counseling in Medical Genetics*, 158.

71. Ibid., 153.

72. I compiled the information about the Heredity Clinic's 111 adoption cases by using the alphabetical disease catalogue to consult all the files classified as "Adoption." I then looked at each separate file, located throughout the 68 boxes of the collection, which often consisted of a single Case Record Sheet but sometimes included larger case files. The Disease Catalogue is included in Box 1, University of Michigan Adult Medical Genetics Clinic Records (AMGCR), BHL, UM. Access to the records was approved by the University of Michigan Institutional Review Board, HUM00012519. Only pseudonyms are used to protect patient confidentiality.

73. Similarly, to determine the approximate number of cases related to adoption filed under "Racial Characteristics," I used the alphabetical Disease Catalogue in Box 1 of the collection to identify and locate all 129 cases. The figure of 54 adoption-related files is very likely an underestimate; it could be as high as 75. However, in many cases the single Case Record Sheets contained too little information to definitively include the case as adoption related. Box 1, and related files, AMGCR, BHL, UM.

74. Herman, *Kinship by Design*.

75. James V. Neel to Mr. Richard L. McCartney, March 7, 1957, Kindred 4943, Box 26, AMGCR, BHL, UM.

76. James V. Neel to Mrs. Marie McCracken, November 12, 1959, Kindred 7408, Box 29 AMGCR, BHL, UM.

77. J. A. Fraser Roberts, "Reginald Ruggles Gates, 1882–1962," *Biographical Memoirs of Fellows of the Royal Society* 10 (1964): 91.

78. Gavin Schaffer, "'Scientific' Racism Again?': Reginald Gates, the *Mankind Quarterly* and the Question of 'Race' in Science after the Second World War," *Journal of American Studies* 41, no. 2 (2007): 253–78.

79. Juvenile Investigation, Official Case, File no. 1458, Kindred 4943, Box 26, AMGCR, BHL, UM.

80. Richard L. McCartney to Dr. James V. Neel, March 1, 1957, Kindred 4943, Box 26, AMGCR, BHL, UM.

81. James V. Neel to Judge Robert P. Polleys, March 14, 1957; and Statement by James V. Neel, March 13, 1957, Kindred 4943, Box 26, AMGCR, BHL, UM.

82. Donald C. Smith to Ernest H. Watson, October 18, 1960, Kindred 7296, Box 29, AMGCR, BHL, UM.

83. E. H. Watson to Donald C. Smith, October 31, 1960, Kindred 7296, Box 29, AMGCR, BHL, UM.

84. Donald C. Smith to Ernest H. Watson, October 18, 1960, Kindred 7296, Box 29, AMGCR, BHL, UM.

85. Margery W. Shaw to Donald C. Smith, November 17, 1961, Kindred 7296, Box 29, AMGCR, BHL, UM.

86. Margery W. Shaw to Mrs. C. Fisher, October 16, 1961; Report of Psychological Examination, September 23, 1961, Kindred 7296, Box 29, AMGCR, BHL, UM.

87. James V. Neel to Mrs. Virginia Jackson, January 23, 1970, Kindred 10882, Box 66, AMGCR, BHL, UM.

88. Verle E. Headings to Bernard Kazyak, June 15, 1967, Kindred 10138, Box 66, AMGCR, BHL, UM.

89. Pearl Buck, *Children for Adoption* (New York: Random House, 1964); Pearl Buck, "Should White Families Adopt Brown Babies?," *Ebony*, June 1958, 26–30.

90. Theresa Monsour, "Retired Couple Share Passion for a Busy Life," *St. Paul Pioneer Press*, June 12, 1990, 1A, 5A, Biographical File: Sheldon Reed (SR), UMA.

91. Danielle Deaver, "WFU Medical School Apologizes Again for Role: Officials Critized in Choice of a Supporter," from Epilogue of "Against Their Will: North Carolina's Sterilization Program" (2002), Web-based supplement, *Winston-Salem Journal*, Journalnow, http://extras.journalnow.com/againsttheirwill/parts/epilogue/print story27.html.

92. Connelly, *Fatal Mis-Conception*.

93. Stern, *Eugenic Nation*.

94. See, for example, Michael M. Kaback and Robert S. Zieger, "Heterozygote Detection in Tay-Sachs Disease: A Prototype Community Screening Program for the Prevention of Recessive Genetic Disorders (1972)," in *Landmarks in Medical Genetics: Classic Papers with Commentaries*, ed. Peter S. Harper (Oxford: Oxford University Press, 2004), 279–84.

95. See Howard Markel, "Scientific Advances and Social Risks: Historical Perspectives of Genetic Screening Programs for Sickle Cell Disease, Tay-Sachs Disease, Neural Tube Defects, and Down Syndrome, 1970–1997," in *Promoting Safe and Effective Genetic Testing in the United States. Final Report of the Task Force on Genetic Screening (Appendix II)*, ed. Neil A. Holtzman and Michael S. Watson (Bethesda, MD: NIH-DOE Working Group on Ethical, Legal and Social Implications of Human Genome Research, 1997), www.genome.gov/10002401; Keith Wailoo and Stephen Pemberton, *The Troubled Dream of Genetic Medicine: Ethnicity and Innovation in Tay-Sachs, Cystic Fibrosis, and Sickle Cell Disease* (Baltimore: Johns Hopkins University Press, 2006); Sherry I. Brandt-Rauf et al., "Ashkenazi Jews and Breast Cancer: The Consequences of Linking Ethnic Identity to Genetic Disease," *American Journal of Public Health* 96, no. 11 (2006): 1979–88; and Duster, *Backdoor to Eugenics*.

96. Harriet A. Washington, *Medical Apartheid: The Dark History of Medical Experimentation and Black Americans from Colonial Times to the Present* (New York: Anchor Books, 2008); Alexandra Minna Stern, "Sterilized in the Name of Public

Health: Race, Immigration, and Reproductive Control in Modern California, *American Journal of Public Health* 95, no. 7 (2005): 1128–38.

97. Nancy Steinberg Warren, interviewed by the author, March 18, 2011.

98. See www.geneticcounselingtoolkit.com/about.htm.

99. Vivian Ota Wang, "Multicultural Genetic Counseling: Then, Now, and in the 21st Century," *American Journal of Medical Genetics* 106, no. 3 (2001): 208–15; Ilana Suez Mittman and Katy Downs, "Diversity in Genetic Counseling: Past, Present and Future," *Journal of Genetic Counseling* 17, no. 4 (2008): 301–13; Stephanie C. Smith, Nancy Steinberg Warren, and Lavanya Misra, "Minority Recruitment into the Genetic Counseling Profession," *Journal of Genetic Counseling* 2, no. 3 (1993): 171–81; Tracey Oh and Linwood J. Lewis, "Consideration of Genetic Counseling as a Career: Implications for Diversifying the Genetic Counseling Field," *Journal of Genetic Counseling* 14, no. 1 (2005): 71–81.

CHAPTER FOUR: *Disability*

1. *The Dark Corner* transcript, Folder: Films, Robert E. Cooke Papers (RC), Alan Mason Chesney Medical Archives (AMCMA), Johns Hopkins Medical Institutions (JHMI).

2. Cri du chat syndrome results from a subtelomeric deletion (5p), and clinical severity is associated with the size of the deletion. The name reflects the catlike cry that is characteristic of this condition. "Submicroscopic Chromosomal Abnormalities and the Chromosomal Phenotype," in Helen V. Firth, Jane A. Hurst, and Judith G. Hall, *Oxford Desk Reference: Clinical Genetics* (Oxford: Oxford University Press, 2005), 548.

3. In a letter to a local mother, Cooke sympathized with the challenges of parenting a retarded child: "I am thoroughly familiar with the problems of raising seriously defective children, having two of my own who were much more retarded than the usual mongoloid child. They lived with us for 20 years, neither being able to talk or walk." Robert E. Cooke to Mrs. Marilyn Birkmeyer, February 27, 1973, RC, AMCMA, JHMI; also see Robert E. Cooke, recorded interview by John F. Stewart, March 29, 1968, John F. Kennedy Library Oral History Program.

4. *The Dark Corner* transcript, 4–5.

5. Ibid., 17.

6. David J. Rothman and Sheila M. Rothman, *The Willowbrook Wars: Bringing the Mentally Disabled into the Community* (New Brunswick, NJ: AldineTransaction, 2005 [1984]); Joel D. Howell and Rodney A. Hayward, "Writing Willowbrook, Reading Willowbrook: The Recounting of a Medical Experiment," in *Useful Bodies: Humans in the Service of Medical Science in the Twentieth Century*, ed. Jordan Goodman, Anthony McElligott, and Lara Marks (Baltimore: Johns Hopkins University Press, 2003), 190–213.

7. James W. Trent Jr., *Inventing the Feeble Mind: A History of Mental Retardation in the United States* (Berkeley: University of California Press, 1994); Allison C. Carey, *On the Margins of Citizenship: Intellectual Disability and Civil Rights in Twentieth-Century America* (Philadelphia: Temple University Press, 2009).

8. Lennard J. Davis, "Constructing Normalcy," in *The Disability Studies Reader*,

ed. Lennard J. Davis, 3rd ed. (New York: Routledge, 2010), 3–192; Erik Parens and Adrienne Asch, eds., *Prenatal Testing and Disability Rights* (Washington, DC: Georgetown University Press, 2000); Douglas C. Baynton, "Disability and the Justification of Inequality in American History," in *The New Disability History: American Perspectives*, ed. Paul K. Longmore and Lauri Umansky (New York: New York University Press, 2001), 33–57; Richard K. Scotch, "'Nothing about Us without Us': Disability Rights in America," *OAH Magazine of History* 23, no. 3 (2009): 17–22.

9. Nathaniel Comfort, *The Science of Human Perfection: Heredity, Health, and Human Improvement in American Biomedicine* (New Haven, CT: Yale University Press, 2012); Diane B. Paul, *The Politics of Heredity: Essays on Eugenics, Biomedicine, and the Nature-Nurture Debate* (Albany: State University of New York Press, 1998); Molly Ladd-Taylor, "'A Kind of Genetic Social Work': Sheldon Reed and the Origins of Genetic Counseling," in *Women, Health, and Nation: Canada and the United States since 1945*, ed. Georgina Feldberg et al. (Montreal: McGill University Press, 2003), 67–83; Ruth Schwartz Cowan, *Heredity and Hope: The Case for Genetic Screening* (Cambridge, MA: Harvard University Press, 2008); Allen Buchanan et al., *From Chance to Choice: Genetics & Justice* (Cambridge: Cambridge University Press, 2000).

10. Michel Foucault's concept of "subjugated knowledges" is applicable to these reconfigured identities. See Michel Foucault, *Power/Knowledge: Selected Interviews and Other Writings 1972–1977*, ed. Colin Gordon (New York: Pantheon, 1980).

11. See Nikolas Rose, *The Politics of Life Itself: Biomedicine, Power, and Subjectivity in the Twenty-First Century* (Princeton, NJ: Princeton University Press, 2006).

12. Reed's first formal publication discussing genetic counseling was Sheldon C. Reed, *Reactivation of the Dight Institute, 1947–1949, Counseling in Human Genetics*, Bulletin no. 6 (Minneapolis: University of Minnesota Press, 1949).

13. Curriculum Vita, Sheldon C. Reed, Biographical File: Sheldon Reed (SR), University of Minnesota Archives (UMA); V. Elving Anderson, "Sheldon C. Reed, Ph.D. (November 7, 1910–February 1, 2003): Genetic Counseling, Behavioral Genetics," *Behavior Genetics* 33, no. 6 (2003): 630, SR, UMA.

14. Ken Der, "Retired Genetics Prof Keeps Himself Busy," *Minnesota Daily*, February 2, 1982, SR, UMA.

15. S. C. Reed to Kenneth M. Ludmerer, July 28, 1970, Box 1, Dight Institute of Human Genetics Research (DIHGR), UMA.

16. S. C. Reed, "From Tree House to Family Tree," Box 4; Robert J. R. Johnson, "Heredity Explained—U Team Calls Toss on Genetic Gamble," *Pioneer Press*, March 16, 1958, n.p.; John Medelman, "The Incredible Dr. Dight: His Crusade to Abolish Wickedness," *Select Twin Citian*, July 1961, 11–13, Box 2, DIHGR, UMA; also see Neal Ross Holtan, "From Eugenics to Public Health Genetics in Mid-Twentieth Century Minnesota" (PhD dissertation, University of Minnesota, 2011).

17. Brochure, Relating to the Minnesota Eugenics Society, n.d. (likely late 1920s), Box 5, DIHGR, UMA.

18. Annual Report of the Dight Institute, 1941–1942, Box 1, Minnesota Human Genetics League Records (MHGLR), UMA; Clarence P. Oliver, "A Report on the Organization and Aims of the Dight Institute," and Evadene Burris Swanson, "A

Biographical Sketch of Charles Fremont Dight," *Dight Institute of the University of Minnesota Bulletin* 1 (1943); "Annual Report of the Dight Institute" (1941–1942), MHGLR, UMA; Dick Margolis, "Dight Studies Show Genes May Decide Your Future," *Minneapolis Daily*, May 27, 1949, n.p., Box 3, DIHGR, UMA.

19. Clarence P. Oliver, "A Report of the Dight Institute for the Year 1943–1944," *Dight Institute of the University of Minnesota Bulletin* 3 (1945): 1.

20. Statement of C. P. Oliver in Lee R. Dice, "A Panel Discussion: Genetic Counseling," *American Journal of Human Genetics* 4, no. 4 (1952): 341.

21. Clarence P. Oliver to the Dight Institute Committee, n.d. (likely early 1940s), Box, 1, MHGLR, UMA.

22. Minutes of the Dight Institute Committee Meeting of September 18, 1945, Box 1, MHGLR, UMA.

23. Minnesota Human Genetics League, Inc., News Bulletin no. 1, November 1945, Box 1; Minnesota Human Genetics League History, 9/1969, Box 1, MHGLR, UMA; Dight Institute Committee Meeting of October 8, 1945, Box 1; Dight Institute Committee Meeting of October 23, 1945, DIHGR, UMA; C. P. Oliver, Report to the Committee of the Dight Institute for the Year 1945–1946, Box 2, DIHGR, UMA.

24. Sheldon C. Reed, "The Local Eugenics Society," *American Journal of Human Genetics* 9, no. 1 (1957): reprint, DIHGR, UMA.

25. Meeting of Board of Directors, Minnesota Human Genetics League, May 14, 1946, Box 1, MHGLR, UMA.

26. S. C. Reed to Mr. Joseph T. Velardo, September 30, 1948, Box 2, DIHGR, UMA.

27. S. C. Reed, Conversational Report to the Dight Committee, November 21, 1947, included in S. C. Reed to Dean T. C. Blegen, November 21, 1947, Box 1, DIHGR, UMA.

28. S. C. Reed to Dr. M. Demerec, December 5, 1947, Box 1, DIHGR, UMA; "University Gets Gift of 40,000 Family Trees," *Minneapolis Daily Times*, April 9, 1948; "40,000 Family Histories—U Gets Valuable Data," most likely from the *Pioneer Press* during the same time period, Box 2, DIHGR, UMA.

29. A. C. Rogers and Maud A. Merrill, *Dwellers in the Vale of Siddem: A True Story of the Social Aspect of Feeble-Mindedness* (Boston: Richard G. Badger, Gorham Press, 1919), 15.

30. See the perceptive analysis of this tract by Molly Ladd-Taylor, "Coping with a 'Public Menace': Eugenic Sterilization in Minnesota," *Minnesota History* 59, no. 6 (2005): 237–48.

31. S. C. Reed to E. J. Engberg, March 14, 1950, Box 1, DIHGR, UMA; The Dight Institute of the University of Minnesota, *Report of Progress, 1949–1951* (Minneapolis: University of Minnesota Press, 1949).

32. D. E. Minnich to Dr. Hastings, mid-December 1951; Preliminary Report on Project E—1126, Between the Division of Public Institutions and the University of Minnesota, "Genetics of Mental Deficiency," Box 1, MHGLR, UMA.

33. Elizabeth W. Reed and Sheldon C. Reed, *Mental Retardation: A Family Study* (a project of the Minnesota Human Genetics League) (Philadelphia: W. B. Saunders, 1965).

34. Ibid., 14.

35. Ibid., 71.

36. Trent Jr., *Inventing the Feeble Mind.*

37. Kathleen W. Jones, "Education for Children with Mental Retardation: Parent Activism, Public Policy, and Family Ideology in the 1950s," in *Mental Retardation in America: A Historical Reader*, ed. Steven Noll and James W. Trent Jr. (New York: New York University Press, 2004), 322–50; Katherine Castles, "'Nice, Average Americans': Postwar Parents' Groups and the Defense of the Normal Family," in *Mental Retardation in America*, ed. Noll and Trent Jr., 351–70.

38. Jane Brody, "Treatise Suggests Ways to Reduce Retardation," *Minneapolis Tribune*, May 2, 1965, 14B, Box 2, DIHGR, UMA.

39. E. W. Reed and S. C. Reed, *Mental Retardation*, 72.

40. S. C. Reed, "Genetic Indications for Therapeutic Abortion," Box 5, DIHGR, UMA.

41. S. C. Reed, "Selected Counseling Cases," Box 2, DIHGR, UMA.

42. S. C. Reed, "Etiology and Genetic Counseling in Mental Retardation" (1955), Box 4, DIHGR, UMA.

43. Israel Zwerling, "Initial Counseling of Parents with Mentally Retarded Children," *Journal of Pediatrics* 44 (1954): 470.

44. Charlotte H. Waskowitz, "The Parents of Retarded Children Speak for Themselves," *Journal of Pediatrics* 54, no. 3 (1959): 322.

45. S. C. Reed, Untitled talk (begins "Thank you, Dr. Jensen," n.d., likely 1960s), likely in Box 4, DIHGR, UMA.

46. "Hereditary Mental Defects Called Curable," *Minneapolis Sunday Tribune*, June 8, 1952, 4, Box 2, DIHGR, UMA.

47. Pearl S. Buck, *The Child Who Never Grew* (Vineland, NJ: Woodbine House, 1992 [1950]); Dale Evans Rogers, *Angel Unaware* (Westwood, NJ: Fleming H. Revell, 1953).

48. Eunice Kennedy Shriver, "Hope for Retarded Children," *Saturday Evening Post*, September 22, 1962, 72. An excellent study of Shriver's dedication to the cause of mental retardation is Edward Shorter, *The Kennedy Family and the Story of Mental Retardation* (Philadelphia: Temple University Press, 2000).

49. Dorothy Garst Murray, *This Is Stevie's Story*, rev. and enlarged ed. (Nashville: Abingdon Press, 1967 [1956]), 62.

50. Ibid., 116–17.

51. Carey, *On the Margins of Citizenship*, 113.

52. Selected Counseling Cases, n.d. (likely 1960s), Box 2, DIHGR, UMA.

53. S. C. Reed, "Counseling for the Parents of the Retarded," Box 4, DIHGR, UMA.

54. Reed, Untitled talk ("Thank you, Dr. Jensen").

55. Mr. and Mrs. Lyle Giesike to S. C. Reed, July 28, 1954, Box 1, DIHGR, UMA.

56. Mildred Thomson, *Prologue: A Minnesota Story of Mental Retardation* (Minneapolis: Gilbert, 1963), 149.

57. Mrs. Miles J. Hubbard to S. C. Reed, January 20, 1962, and "No Simple Answer to Retardation, Genetics Expert Tells County Group," Box 2, DIHGR, UMA.

58. Mrs. Floyd Frank to S. C. Reed, October 21, 1963, Box 2, DIHGR, UMA.

59. S. C. Reed to Elizabeth M. Boggs, April 20, 1961, Box 1, DIHGR, UMA. Also see United States. President's Panel on Mental Retardation, *A Proposed Program for National Action to Combat Mental Retardation* (Washington, DC: Government Printing Office, 1962); and Elizabeth M. Boggs, recorded interview by John F. Stewart, July 17, 1968, John F. Kennedy Library Oral History Program.

60. S. C. Reed, "Genetic Indications for Therapeutic Abortion" (1971), Box 5, DIHGR, UMA.

61. As described and quoted in Marc Lappé, "Can Eugenic Policy Be Just?," chap. 21 in *The Prevention of Genetic Disease and Mental Retardation*, ed. Aubrey Milunsky (Philadelphia: W. B. Saunders, 1975), 465.

62. Memorandum, A. B. Rosenfeld, MD, to Members of the Minnesota Advisory Committee on Human Genetics, May 6, 1960, Box 1; S. C. Reed to Representative William Shovell, March 8, 1961, Box 2; Brochure, Human Genetics, A New Service for Minnesota, early 1960s, Box 2, DIHGR, UMA.

63. Paul Rabinow, "Artificiality and Enlightenment: From Sociobiology to Biosociality," chap. 5 in *Essays on the Anthropology of Reason* (Princeton, NJ: Princeton University Press, 1996), 91–111; also see Kaja Finkler, *Experiencing the New Genetics: Family and Kinship on the Medical Frontier* (Philadelphia: University of Pennsylvania Press, 2000); and Aviad E. Raz, *Community Genetics and Genetic Alliances: Eugenics, Carrier Testing, and Networks of Risk* (New York: Routledge, 2009).

64. Foucault, *Power/Knowledge*.

65. See David S. Newberger, "Down Syndrome: Prenatal Risk Assessment and Diagnosis," *American Family Physician*, August 15, 2009, www.aafp.org/afp/2000 0815/825.html.

66. David Wright, "Mongols in Our Midst: John Langdon Down and the Ethnic Classification of Idiocy, 1858–1924," in *Mental Retardation in America*, ed. Noll and Trent Jr., 92–119.

67. Daniel J. Kevles, "'Mongolian Idiocy': Race and Its Rejection in the Understanding of Mental Disease," in *Mental Retardation in America*, ed. Noll and Trent Jr., 120–29.

68. Quoted in David Goode, *"And Now Let's Build a Better World": The Story of the Association for the Help of Retarded Children, New York City, 1948–1998*, 18. Available at www.ahrcnyc.org.

69. Wolf Wolfensberger and Richard A. Kurtz, "Use of Retardation-Related Diagnostic and Descriptive Labels by Parents of Retarded Children," *Journal of Special Education* 8, no. 2 (1974): 131–42.

70. Quoted in Carey, *On the Margins of Citizenship*, 156.

71. See Adrienne Asch, "Disability and Genetics: A Disability Rights Perspective," *Encyclopedia of the Life Sciences* (2006), an online resource available at http://onlinelibrary.wiley.com/doi/10.1038/npg.els.0005212/full; also see Jackie Leach Scully, "Disability and Genetics in the Era of Genomic Medicine," *Nature Reviews Genetics* 9 (2008): 797–99.

72. First Draft, Statement, "Mongolism," included with Letter to Colleagues, December 12, 1960, Down's Syndrome / Mongolism Folder, MS Coll 5, Papers of Curt

Stern (CS), American Philosophical Society Library (APSL), Philadelphia, Pennsylvania.

73. Hideo Nishimura to Stern, December 28, 1960, Down's Syndrome / Mongolism Folder, CS, APSL.

74. Letter to Colleagues, December 12, 1960, Down's Syndrome / Mongolism Folder, CS, APSL.

75. Irene Uchida to Curt Stern, January 5, 1960, Down's Syndrome / Mongolism Folder, CS, APSL. This collective statement appeared as "Mongolism," *Lancet*, April 8, 1961, 775.

76. See Fiona Alice Miller, "Dermatoglyphics and the Persistence of 'Mongolism': Networks of Technology, Disease and Discipline," *Social Studies of Science* 33, no. 1 (2003): 75–94; also see Fiona Alice Miller et al., "Redefining Disease? The Nosological Implications of Molecular Genetic Knowledge," *Perspectives in Biology and Medicine* 49, no. 1 (2006): 99–114.

77. David Hendin and Joan Marks, *The Genetic Connection: How to Protect Your Family against Hereditary Disease* (New York: William Morrow, 1978), 74.

78. Martha M. Jablow to Ms. Marks, October 25, 1978, Record Group 2-Marks, Joan-Speeches, Lectures and Writings, Human Genetics Graduate Program Records (HGGPR), Sarah Lawrence College Archives (SLCA), Bronxville, New York.

79. Joan H. Marks to Martha M. Jablow, November 9, 1978, Record Group 2-Marks, Joan-Speeches, Lectures and Writings, HGGPR, SLCA.

80. See, for example, Jason Kingsley and Mitchell Levitz, *Count Us In: Growing Up with Down Syndrome* (Orlando: Harvest Book, 2007 [1994]); William I. Cohen, Lynn Nadel, and Myra E. Madnick, *Down Syndrome: Visions for the 21st Century* (New York: Wiley-Liss, 2002).

81. Campbell K. Brasington, "What I Wish I Knew Then . . . Reflections from Personal Experiences in Counseling about Down Syndrome," *Journal of Genetic Counseling* 16 (2007): 731.

82. Ibid.

83. Judith Tsipis, interviewed by the author, August 5, 2010.

84. Annette Kennedy, interviewed by the author, August 30, 2011.

85. Judith Tsipis, e-mail message to author, July 29, 2011.

86. Wendy Uhlmann, interviewed by the author, April 23, 2007, and September 8, 2010.

87. Ibid.; Matt Schudel, "Frank Uhlmann: Psychologist Wrote, Spoke on Living with MS," *Washington Post*, June 29, 2008, C08.

88. Robin Bennett, interviewed by the author, February 13, 2008.

89. "Angelman Syndrome," in Firth, Hurst, and Hall, *Clinical Genetics*, 272–73.

90. Bennett, interview.

91. Catherine A. Reiser and Joan Burns, interviewed by the author, October 22, 2010.

92. Joan Burns, interviewed by Peter Harper, May 12, 2003, www.genmedhist .info/Interviews%20/Burns.

93. Anna Middleton, J. Hewison, and R. F. Mueller, "Attitudes of Deaf Adults toward Genetic Testing for Hereditary Diseases," *American Journal of Human Genetics* 63 (1998): 1175–80.

94. Katy Downs, interviewed by the author, March 30, 2007; Ilana Suez Mittman and Katy Downs, "Diversity in Genetic Counseling: Past, Present and Future," *Journal of Genetic Counseling* 17 (2008): 301–13.

95. See Aubrey Milunsky, ed., *The Prevention of Genetic Disease and Mental Retardation* (Philadelphia: W. B. Saunders, 1975).

96. See Erik Parens and Adrienne Asch, "The Disability Rights Critique of Prenatal Genetic Testing," Special Supplement, *Hastings Center Report* 29, no. 5 (1999): S1–S22; Adrienne Asch, "Prenatal Diagnosis and Selective Abortion: A Challenge to Practice and Policy," *American Journal of Public Health* 89, no. 11 (1999): 1649–57; and Sonia Mateu Suter, "The Routinization of Prenatal Testing," *American Journal of Law and Medicine* 28 (2002): 233–70.

97. See Norma J. Waitzman, Richard M. Scheffler, and Patrick S. Romano, *The Cost of Birth Defects: Estimates of the Value of Prevention* (Lanham, MD: University Press of America, 1996).

98. See Tom Shakespeare, *Disability Rights and Wrongs* (New York: Routledge, 2006).

99. Ibid., 100. An attempt to productively navigate this tension is William M. McMahon, Bonnie Jeanne Baty, and Jeffery Botkin, "Genetic Counseling and Ethical Issues for Autism," *American Journal of Medical Genetics Part C* 142C (2006): 52–57; also see Rayna Rapp, *Testing Women, Testing the Fetus: The Social Impact of Amniocentesis in America* (New York: Routledge, 2000).

100. Linda L. McCabe and Edward R. B. McCabe, "Down Syndrome: Coercion and Eugenics," *Genetics in Medicine* 13, no. 8 (2011): 710.

101. See Brian Skotko, "Mothers of Children with Down Syndrome Reflect on Their Postnatal Support," *Pediatrics* 115, no. 1 (2005): 64–77; and Brian Skotko, "Prenatally Diagnosed Down Syndrome: Mothers Who Continued Their Pregnancies Evaluate Their Health Care Providers," *American Journal of Obstetrics and Gynecology* 192 (2005): 670–77.

102. See Darrin P. Dixon, "Informed Consent or Institutionalized Eugenics? How the Medical Profession Encourages Abortion of Fetuses with Down Syndrome," *Issues in Law & Medicine* 24, no. 1 (2008): 3–59.

103. See Tamsen M. Caruso, Marie-Noel Westgate, and Lewis B. Holmes, "Impact of Prenatal Screening on the Birth Status of Fetuses with Down Syndrome at an Urban Hospital, 1972–1994," *Genetics in Medicine* 1, no. 1 (1998): 22–28; Peter A. Benn et al., "The Centralized Prenatal Genetics Program of New York City III: The First 7,000 Cases," *American Journal of Medical Genetics* 20 (1985): 369–84; Ralph L. Kramer et al., "Determinants of Parental Decisions after the Prenatal Diagnosis of Down Syndrome," *American Journal of Medical Genetics* 79 (1998): 172–74; Mark I. Evans et al., "Prenatal Decisions to Terminate/Continue Following Abnormal Cytogenetic Prenatal Diagnosis: 'What' Is Still More Important Than 'When,'" *American Journal of Medical Genetics* 61 (1996): 353–55; Victoria A. Vincent et al., "Pregnancy Termination because of Chromosomal Abnormalities: A Study of 26,950 Amniocenteses in the Southeast," *Southern Medical Journal* 84, no. 10 (1991): 1210–13; Arie Drugan et al., "Determinants of Parental Decisions to Abort for Chromosome Abnormalities," *Prenatal Diagnosis* 10 (1990): 483–90; Marion S. Verp et al., "Parental

Decision Following Prenatal Diagnosis of Fetal Chromosome Abnormality," *American Journal of Medical Genetics* 29 (1988): 613–22.

104. In two California counties monitored by the California Birth Defects Monitoring Program from 1989 to 1991, a 46.3% decrease in Down syndrome life births was calculated for whites, while the rate for Hispanics was only 10%, demonstrating much lower rates of pregnancy termination for Down syndrome among Hispanics. See Jennifer Bishop et al., "Epidemiologic Study of Down Syndrome in a Racially Diverse California Population, 1989–1991," *American Journal of Epidemiology* 145, no. 2 (1997): 134–47.

105. Brian L. Shaffer, Aaron B. Caughey, and Mary E. Norton, "Variation in the Decision to Terminate Pregnancy in the Setting of Fetal Aneuploidy," *Prenatal Diagnosis* 26 (2006): 667–71.

106. See Csaba Siffel et al., "Prenatal Diagnosis, Pregnancy Terminations and Prevalence of Down Syndrome in Atlanta," *Birth Defects Research (Part A)* 7 (2004): 565–671. Studies have also shown that African American women and Latinas are less likely to undergo prenatal diagnosis. See Miriam Kupperman, Elena Gates, and A. Eugene Washington, "Racial-Ethnic Differences in Prenatal Diagnostic Test Use and Outcomes: Preferences, Socioeconomics, or Patient Knowledge?," *Obstretrics & Gynecology* 87, no. 5, Part 1 (1996): 675–82.

107. "Screening for Fetal Chromosomal Abnormalities," ACOG Practice Bulletin, 77 (January 2007), published in *Obstetrics & Gynecology* 109, no. 1 (2007): 217–27.

108. Patricia E. Bauer, "What's Lost in Prenatal Testing," *Washington Post*, January 14, 2007, B07.

109. George F. Will, "Golly, What *Did* Jon Do?," *Newsweek*, January 29, 2007, www.thedailybeast.com/newsweek/2007/01/28/golly-what-did-jon-do.html; also see his "Eugenics by Abortion: Is Perfection an Entitlement?," *Washington Post*, April 24, 2005, A27.

110. Janice Edwards, interviewed by the author, August 27, 2010; "Toward Concurrence: Understanding Prenatal Screening and Diagnosis of Down Syndrome from the Health Professional and Advocacy Community Perspectives," June 17, 2009, www.ndsccenter.org/news/documents/ConsensusConversationStatement.pdf.

111. This and all the NSGC's position statements can be viewed at www.nsgc.org/Advocacy/PositionStatements/tabid/107/Default.aspx.

112. Anne C. Madeo et al., "The Relationship between the Genetic Counseling Profession and the Disability Community: A Commentary," *American Journal of Medical Genetics Part A* 155, no. 8 (2011): 1777–85.

113. Ibid., 1779.

114. See Jan Hodgson and Jon Weil, "Talking about Disability in Prenatal Genetic Counseling: A Report of Two Interactive Workshops," *Journal of Genetic Counseling* 21, no. 1 (2012): 17–23; and Robert Resta, "Are Genetic Counselors Just Misunderstood? Thoughts on 'The Relationship between the Genetic Counseling Profession and the Disability Community: A Commentary,'" *American Journal of Medical Genetics Part A* 155, no. 8 (2011): 1786–87.

1. Melissa Richter quoted in "Degrees Offered in Genetic Counseling," *New York Times*, December 6, 1970, 71.

2. Regina H. Kenen, "Opportunities and Impediments for a Consolidating and Expanding Profession: Genetic Counseling in the United States," *Social Science & Medicine* 45, no. 9 (1997): 1377–86; also see the ABGC website (www.abgc.net/english/view.asp?x=1) and the NSGC website (www.nsgc.org); Audrey Heimler, "An Oral History of the National Society of Genetic Counselors," *Journal of Genetic Counseling* 6, no. 3 (1997): 315–36.

3. Robin Morgan, ed., *Sisterhood Is Powerful: An Anthology of Writings from the Women's Liberation Movement* (New York: Vintage, 1970).

4. Terry H. Anderson, *The Movement and the Sixties* (Oxford: Oxford University Press, 1996).

5. Edith Evans Asbury, "Sarah Lawrence Bypassing a Sit-In," *New York Times*, March 6, 1969, 26; "Sarah Lawrence Quiet," *New York Times*, March 15, 1969, 22; Thomas F. Brady, "Mrs. Raushenbush Emerges Unscarred in Sarah Lawrence Confrontation," *New York Times*, March 23, 1969, 70; "Sarah Lawrence Institute Center of Student Protest," *New York Times*, May 13, 1969, 33.

6. Melissa Richter to Mrs. Esther Raushenbush, October 6, 1969, Record Group (RG) 2, Human Genetics Graduate Program Records (HGGPR), Sarah Lawrence College Archives (SLCA), Bronxville, New York.

7. M. Richter to Sheldon C. Reed, April 23, 1969, RG 2, HGGPR, SLCA.

8. Melissa L. Richter and Jane Banks Whipple, *A Revolution in the Education of Women: Ten Years of Continuing Education at Sarah Lawrence College* (Bronxville, NY: Sarah Lawrence College, 1972).

9. Cited in ibid., 7.

10. Richter, "Health Sciences in Social Change: Pilot Projects at Sarah Lawrence College," RG 2, HGGPR, SLCA.

11. Marylin Bender, "A New Breed of Middle-Class Women Emerging," *New York Times*, March 17, 1969, 34; Ruth Rosen, *The World Split Open: How the Women's Movement Changed America* (New York: Penguin, 2000).

12. David Harris et al., "Legal Abortion 1970–1971—the New York City Experience," *American Journal of Public Health* 63, no. 5 (1973): 409–18; James C. Mohr, *Abortion in America: The Evolution of National Policy* (Oxford: Oxford University Press, 1979).

13. Wendy Kline, *Bodies of Knowledge: Sexuality, Reproduction, and Women's Health in the Second Wave* (Chicago: University of Chicago Press, 2010).

14. The National Foundation—March of Dimes, "Genetic Counseling with Particular Reference to Anticipatory Guidance and the Prevention of Birth Defects," *Birth Defects: Original Article Series* 6, no. 1 (Baltimore: Williams and Wilkins, 1970); Richter, "Notes on Symposium on Genetic Counseling," January 29, 1969, RG 2, HGGPR, SLCA; Robert W. Stock, "Will the Baby Be Normal? The Genetic Counselor Tries to Find the Answer by Translating the Biological Revolution into Human Terms," *New York Times* (Sunday magazine), March 23, 1969, 25–27, 79, 82–84, 94–96.

15. Audrey Heimler, interviewed by the author, October 13, 2007.

16. M. Richter, "History, Human Genetics Program, Sarah Lawrence College," RG 9, HGGPR, SLCA.

17. "Memorial Dec. 21 for Dr. Richter," *Providence Journal*, December 5, 1974, B2.

18. Academic Personnel File of Melissa Richter, SLCA.

19. Melissa Richter, Vita, Richter Biographical Files, ca. 1970–1976, HGGPR, SLCA.

20. Jacquelyn Mattfeld, interviewed by the author, April 17, 2008.

21. "Happenings," *Sarah Lawrence Alumnae Magazine*, Spring 1971, 5.

22. Frank A. Walker to Mrs. Melissa L. Richter, March 28, 1969; Jeanette Schulz to Mrs. Melissa Richter, March 25, 1969; Stanley W. Wright to Mrs. Melissa Richter, April 2, 1969, RG 2, HGGPR, SLCA.

23. Sheldon C. Reed to Melissa L. Richter, April 2, 1969, RG 2, HGGPR, SLCA.

24. See, for example, Melissa L. Richter to Dr. John W. Littlefield, July 3, 1970, RG 3; and Richmond S. Paine to Mrs. Melissa L. Richter, April 7, 1969, RG 2, HGGPR, SLCA.

25. Richter, "Health Sciences in Social Change: Pilot Projects at Sarah Lawrence College," RG 2, HGGPR, SLCA.

26. Melissa Richter, "The Effects of Over-Population on Behavior: The Biologist's View," *Sarah Lawrence Alumnae Magazine*, Spring 1968, 12; Paul R. Ehrlich, *The Population Bomb* (New York: Sierra Club–Ballantine Books, 1968).

27. Nathaniel Comfort, *The Science of Human Perfection: Heredity, Health, and Human Improvement in American Biomedicine* (New Haven, CT: Yale University Press, 2012).

28. M. Richter, "A Talk to the New England Association of Nurses at Boston College," RG 2, HGGPR, SLCA.

29. Steven R. Coleman, "To Promote Creativity, Community, and Democracy: The Progressive Colleges of the 1920s and the 1930s" (doctoral dissertation, Columbia University, 2000); Suzanne Walters, "An Individual Education: The Foundations of Sarah Lawrence College," *Westchester Historian* 79, no. 4 (2003): 100–112.

30. Joan H. Marks, interviewed by the author, January 19, 2007.

31. Jessica Davis, interviewed by the author, March 14, 2008; Mattfeld, interview.

32. Marks, interview.

33. Jacquelyn Mattfeld to Mr. Quigg Newton, December 20, 1968, RG 1, HGGPR, SLCA.

34. Grant Awards to the Human Genetics Program, 1969–1992, RG 1, HGGPR, SLCA.

35. Davis, interview.

36. Ellie Miller to Human Genetics Students, August 18, 1972, RG 6, HGGPR, SLCA; Mattfeld, interview.

37. Jessica G. Davis to Dr. John B. Graham, December 18, 1972, RG 2, HGGPR, SLCA.

38. Sheldon C. Reed, *Counseling in Medical Genetics* (Philadelphia: W. B. Saunders. 1955).

39. See extensive correspondence between Richter and Hirschhorn, including notes from meetings, in RG 3, HGGPR, SLCA; Kurt Hirschhorn, interviewed by the

author, March 8, 2008; also see Discussion Forum, "A Question of Genes: Inherited Risk," interview with Lynn Godmilow, 1997, www.backbonemedia.org/genes/career/635_godmilow.html.

40. Davis, interview.

41. Graduate Program in Human Genetics, Sarah Lawrence College, Center for Continuing Education, 1969, RG 6, HGGPR, SLCA.

42. John R. Whittier to Melissa Richter, March 20, 1969; and Richter, Notes from meetings and conversations with Whittier, December 10, 1969; February 2, 1970; July 10, 1970; October 22, 1970; and October 26, 1970, RG 2, HGGPR, SLCA.

43. Melissa L. Richter to Henry L. Nadler, April 1, 1971, RG 6, HGGPR, SLCA.

44. Melissa Richter to Curriculum Committee, May 25, 1970, RG 6, HGGPR, SLCA.

45. E-mail correspondence with Audrey Heimler, June 18, 2008; "Details of the Proposed Program," RG 1, HGGPR, SLCA.

46. Human Genetics Program Description of Jobs Held by Graduates, October 1973, RG 1, HGGPR, SLCA.

47. Davis, interview; Motulsky, who eventually changed his attitude, initially was opposed to Richter's plan, telling her, "I would, therefore, strongly dissuade you to set up a training program for genetic counselors. Persons trained under such a program would probably do more harm than good." See Arno G. Motulsky to Mrs. Melissa L. Richter, April 9, 1969, RG 3, HGGPR, SLCA.

48. Mattfeld, interview.

49. Heimler, interview.

50. A defining moment for Marks came in 1974 when she was a plenary speaker at the American Society for Human Genetics annual meeting and presented an overview of Sarah Lawrence's program and discussed the training and future roles of genetic counselors. See Joan H. Marks, "Training of Genetic Associates: A Five Year Report," RG 2, HGGPR, SLCA; and "Clinical Placements—since 1973–1974, Human Genetics Program, December 1977," RG 6, HGGPR, SLCA; Joan H. Marks, e-mail message to author, August 23, 2011.

51. Arno G. Motulsky, "2003 ASHG Award for Excellence in Human Genetics Education: Introductory Speech for Joan Marks," *American Journal of Human Genetics* 74 (2004): 393–94.

52. Joan H. Marks, "Caring for the Whole Patient: Health Advocacy," *Connecticut Medicine* 45, no. 2 (1981): 103–6, RG 2, HGGPR, SLCA.

53. Melissa Richter to Andy (no last name), July 9, 1969, RG 1, HGGPR, SLCA.

54. Richter, "History, Human Genetics Program," Fall 1970, with additions 6/71.

55. Heimler, interview.

56. Joan H. Marks and Melissa Richter, "The Genetic Associate: A New Health Professional," *American Journal of Public Health* 66, no. 4 (1976): 388–90.

57. "Report of a Site Visit of the Advisory Committee to the Human Genetics Program of Sarah Lawrence College" (1973), RG 3, HGGPR, SLCA.

58. Joan H. Marks to Dr. John Opitz, May 1, 1975, RG 2; Joan H. Marks, Memorandum, August 1974, RG 1, HGGPR, SLCA; E-mail correspondence with Joan H. Marks.

59. Changing Student Population, December 1977, RG 3; "Sarah Lawrence Col-

lege Human Genetics Graduate Program," December 1977, and "Report to the President on the Human Genetics Program, 1973–1974," June 1974, RG 1, HGGPR, SLCA.

60. Changing Student Population, December 1977, Administrative Files-Statistics-1969–1979, HGGPR, SLCA.

61. Caroline Lieber, interviewed by the author, November 5, 2010.

62. Asilomar Conference, 1975, RG 2, HGGPR, SLCA; also see A. P. Walker et al., "Report of the 1989 Asilomar Meeting on Education in Genetic Counseling," *American Journal of Human Genetics* 26 (1990): 1223–30.

63. Biographical Information Sheet, July 1965; "Dr. Charlotte Avers, 63, Rutgers professor," *Star-Ledger*, March 14, 1990, n.p., Charlotte Avers Papers, Special Collections and University Archives (SCUA), Rutgers University Archives (RUA).

64. Charlotte J. Avers, "Proposal for a Graduate Program in Human Genetics and Genetic Counseling," April 1971, Douglass College Office of the Dean Records (DCODR), Record Group (RG) 19/AO/02, Box 3, Folder 1, SCUA, RUA.

65. Marian L. Rivas, interviewed by the author, September 28, 2010.

66. Graduate Program in Genetic Counseling, July 1, 1976, DCODR RG 19/AO/02, Box 3, Folder 2, SCUA, RUA.

67. Bulletins of the Graduate School, 1971–1972, and 1972–1973, Course Catalog Collection, SCUA, RUA.

68. Marian Rivas, e-mail message to author, August 19, 2011.

69. Gary Frohlich, Michael Begleiter, June Peters, and Bonnie Baty, interviewed by the author, December 18, 2007.

70. Ibid.

71. Ibid.; Rivas, interview; also see Kenneth W. Fisher to Dean Margery S. Foster, February 26, 1975, on dissatisfaction over limited faculty lines for the program during its early years despite the professional accomplishments of the first several cohorts of graduates, DCODR, RG 19/AO/02, Box 3, Folder 2, SCUA, RUA.

72. *Health and Medical Sciences Program 1977/78* (Berkeley: University of California, Berkeley, September 1977), 4.

73. Leonard J. Duhl and Stephen R. Blum, "The Berkeley Program in Health and Medical Sciences: Origins, Life History, and Aspirations," 1976, Materials related to the Genetic Counseling Program (GCP), University of California at Berkeley (UCB), Personal Archive of Margie Goldstein (MG).

74. Descriptive Materials regarding the Genetic Counseling Option for the Committee on Educational Policy, May 25, 1977, GCP, UCB, MG.

75. Roberta Palmour, interviewed by the author, June 10, 2010.

76. Lucille Poskanzer, interviewed by the author, November 23, 2007.

77. Charles Epstein, interviewed by the author, May 15, 2008; Jon Weil, interviewed by the author, November 20, 2007.

78. Bryan D. Hall, interviewed by the author, August 11, 2010.

79. Palmour, interview.

80. Lucille Poskanzer, e-mail message to author, August 2, 2011.

81. Anne Matthews, interviewed by the author, August 18, 2010.

82. Vincent M. Riccardi, "Regional Genetic Counseling Programs," chap. 18 in *The Prevention of Genetic Disease and Mental Retardation*, ed. Aubrey Milunsky (Philadelphia: W. B. Saunders, 1975), 410–21; Vincent M. Riccardi, "Commentary: Com-

munity Response to a Regional Genetic Counseling Program," in *Genetic Counseling*, ed. Herbert A. Lubs and Felix de la Cruz (New York: Raven Press, 1977), 93–96; Vikki Porter, "How Are Your Genes? It's All Mom and Dad's Fault," *Colorado Springs Sun, Silhouette*, April 1, 1973, 6–7, Box 3, Series 3: Birth Defects (S3), Medical Program Records (MPR), March of Dimes Archives (MDA), White Plains, New York.

83. Matthews, interview.

84. Robin Grubs, interviewed by the author, June 14, 2010.

85. Ann Walker, interviewed by the author, October 13, 2007.

86. Heimler, interview.

87. Frohlich et al., interview.

88. Lieber, interview.

89. Frohlich et al., interview.

90. Barbara Biesecker, interviewed by the author, February 1, 2007.

91. Bonnie LeRoy, interviewed by the author, January 30, 2008.

92. Catherine Reiser and Joan Burns, interviewed by the author, October 22, 2010.

93. Debra Lochner Doyle, interviewed by the author, January 11, 2008.

94. Carol Norem, interviewed by the author, April 30, 2007.

95. Diane Baker, interviewed by the author, January 16, 2008.

96. Amy S. Wharton, "The Sociology of Emotional Labor," *Annual Review of Sociology* 35 (2009): 147–65; Amy S. Wharton and Rebecca J. Erickson, "The Consequences of Caring: Exploring the Links between Women's Job and Family Emotion Work," *Sociological Quarterly* 36, no. 2 (1995): 273–96.

97. Arlie R. Hochschild, *The Managed Heart: The Commercialization of Human Feeling* (Berkeley: University of California Press, 1983); Amy S. Wharton, "The Sociology of Emotional Labor," *Annual Reviews of Sociology* 35 (2009): 147–65.

98. June A. Peters, "Genetic Counselors: Caring Mindfully for Ourselves," chap. 13 in *Genetic Counseling Practice: Advanced Concepts and Skills*, ed. Bonnie S. Leroy, Patricia McCarthy Veach, and Dianne M. Bartels (New York: Wiley-Blackwell, 2010), 307–52; also see Céleste M. Brotheridge and Alicia A. Grandey, "Emotional Labor and Burnout: Comparing Two Perspectives of 'People Work,'" *Journal of Vocational Behavior* 60 (2002): 17–39.

CHAPTER SIX: *Ethics*

1. Daniel Callahan, "The Hastings Center and the Early Years of Bioethics," *Kennedy Institute of Ethics Journal* 9, no. 1 (1999): 56.

2. Ibid., 55.

3. *The Hastings Center Institute of Society, Ethics, and the Life Sciences Studies* 1, no. 1 (1973); Callahan, "Hastings Center and Early Years."

4. See Albert R. Jonsen, *Birth of Bioethics* (Oxford: Oxford University Press, 2003), esp. chap. 6. The Hastings Center's five initial focus areas were organ transplantation, human experimentation, prenatal diagnosis of genetic disease, prolongation of life, and behavior control.

5. Walter G. Peter III, "Ethical Perspectives in the Use of Genetic Knowledge," *BioScience* 21, no. 22 (1971): 1133–37.

6. Daniel Callahan, "Ethics, Law, and Genetic Counseling," *Science* 176 (April 14, 1972): 199.

7. Thomas H. Murray, "Deciphering Genetics," *Hastings Center Report* 39, no. 3 (2009): 19–22.

8. Marc Lappé, "Allegiances of Human Geneticists: A Preliminary Typology," *Hastings Center Studies* 1, no. 2 (1973): 65.

9. Marc Lappé et al., "The Genetic Counselor: Responsible to Whom?," *Hastings Center Studies* 1, no. 2 (1971): 6.

10. Robert G. Resta, "Complicated Shadows: A Critique of Autonomy in Genetic Counseling Practice," in *Genetic Counseling Advanced Practice Text*, ed. Bonnie S. LeRoy, Patricia McCarthy Veach, and Diane B. Bartels (New York: John Wiley and Sons, 2010), 13–30.

11. Ibid.

12. For an impassioned and informed analysis of this difficult question see Robert M. Veatch, *Patient, Heal Thyself: How the NEW MEDICINE Puts the Patient in Charge* (Oxford: Oxford University Press, 2009).

13. See James R. Sorenson, "Genetic Counseling: Values That Have Mattered"; Beth A. Fine, "The Evolution of Nondirectiveness in Genetic Counseling and Implications of the Human Genome Project"; and Arthur L. Caplan, "Neutrality Is Not Morality: The Ethics of Genetic Counseling," in *Prescribing Our Future: Ethical Challenges in Genetic Counseling*, ed. Dianne M. Bartels, Bonnie S. LeRoy, and Arthur L. Caplan (New York: Aldine de Gruyter, 1993), 3–14, 101–18, 149–68.

14. Jon Weil et al., "The Relationship of Nondirectiveness to Genetic Counseling: Report of a Workshop at the 2003 Annual Education Conference," *Journal of Genetic Counseling* 15 (2006): 85–93; Clare Williams, Priscilla Alderson, and Bobbie Farsides, "Is Nondirectiveness Possible within the Context of Antenatal Screening and Testing?," *Social Science & Medicine* 54, no. 3 (2002): 339–47.

15. Jon Weil, "Psychosocial Genetic Counseling in the Post-Nondirective Era: A Point of View," *Journal of Genetic Counseling* 12, no. 3 (2003): 199–211; Barbara Bowles Biesecker, "Back to the Future of Genetic Counseling: Commentary on 'Psychosocial Genetic Counseling in the Post-Nondirective Era,'" *Journal of Genetic Counseling* 12, no. 3 (2003): 213–17.

16. Ann P. Walker, "The Practice of Genetic Counseling," chap. 1 in *A Guide to Genetic Counseling*, ed. Wendy R. Uhlmann, Jane L. Schuette, and Beverly M. Yashar, 2nd ed. (Hoboken, NJ: John Wiley, 2009), 1–36.

17. See Jonsen, *Birth of Bioethics*.

18. Howard Kirschenbaum, *The Life and Work of Carl Rogers* (Ross-on-Wye, UK: PCCS Books, 2007).

19. Ibid., 79; for a discussion of Rogers's philosophy of psychotherapy see Carl R. Rogers, *Client-Centered Therapy: Its Current Practice, Implications, and Theory* (Boston: Houghton Mifflin, 1951).

20. Nathaniel J. Raskin, "The Development of Nondirective Therapy," *Journal of Consulting Psychology* 12 (1948): 92–110.

21. Kirschenbaum, *Life and Work*, 90.

22. Carl R. Rogers, "The Use of Electrically Recorded Interviews in Improving Psychotherapeutic Techniques," *Journal of Orthopsychiatry* 12, no. 3 (1942): 429–34.

23. William U. Snyder, ed., *Casebook of Non-directive Counseling* (New York: Houghton Mifflin, 1947).

24. Carl Rogers and David E. Russell, *Carl Rogers: The Quiet Revolutionary, An Oral History* (Roseville, CA: Penmarin Books, 2002), 245.

25. Ibid.

26. Rogers, *Client-Centered Therapy*; James H. Capshew, *Psychologists on the March: Science, Practice, and Professional Identity in America, 1929–1969* (Cambridge: Cambridge University Press, 1999).

27. Kirschenbaum, *Life and Work*.

28. Ellen Herman, *The Romance of American Psychology: Political Culture in the Ages of Experts* (Berkeley: University of California Press, 1995), esp. chap. 9.

29. See chap. 7 in Ruth R. Faden, Tom L. Beauchamp, and Nancy M. P. King, *A History and Theory of Informed Consent* (Oxford: Oxford University Press, 1986); also see Matthew Clayton, "Individual Autonomy and Genetic Choice," chap. 14 in *A Companion to Genethics*, ed. Justine Burley and John Harris (Malden, MA: Blackwell, 2002), 191–205.

30. See Sheldon C. Reed, "A Short History of Genetic Counseling," *Dight Institute for Human Genetics at the University of Minnesota Bulletin* 14 (1974): 5, which was reprinted in *Social Biology* (formerly *Eugenics Quarterly*) 12, no. 4 (1974): 332–39; also see Robert G. Resta, "In Memoriam: Sheldon Clark Reed, PhD, 1910–2003," *Journal of Genetic Counseling* 12, no. 3 (2003): 283–85.

31. Diane B. Paul, *The Politics of Heredity: Essays on Eugenics, Biomedicine, and the Nature-Nurture Debate* (Albany: State University of New York Press, 1998); Molly Ladd-Taylor, "'A Kind of Genetic Social Work': Sheldon Reed and the Origins of Genetic Counseling," in *Women, Health, and Nation: Canada and the United States since 1945*, ed. Georgina Feldberg et al. (Montreal: McGill University Press, 2003), 67–83.

32. Sheldon C. Reed, *Counseling in Medical Genetics* (Philadelphia: W. B. Saunders, 1955), 11–12.

33. Statement by Sheldon C. Reed in Lee R. Dice, "A Panel Discussion: Genetic Counseling," *American Journal of Human Genetics* 4, no. 4 (1952): 339.

34. Sheldon C. Reed, "Genetic Counseling," *Proceedings of Symposium on Human Genetics in Public Health* (University of Minnesota, August 9–11, 1964), 36. I thank Debra Doyle for sending me a PDF of this volume.

35. Sheldon C. Reed, Untitled talk (begins "Thank you, Dr. Jensen," n.d., likely 1960s), likely in Box 4, Dight Institute for Human Genetics Records (DIHGR), University of Minnesota Archives (UMA).

36. S. C. Reed, "Genetic Counseling (Women's Studies)" (1997), Box 3, DIHGR, UMA.

37. Statement of C. P. Oliver in Dice, "Panel Discussion," 341, 342, 343.

38. Oral History Interview with F. Clarke Fraser, conducted by Andrea Maestrejuan, October 27–28, 2004, Oral History of Human Genetics Project, UCLA and Johns Hopkins University, http://ohhgp.pendari.com/Links.aspx.

39. S. C. Reed to Dr. Harry L. Shapiro, May 15, 1961; S. C. Reed to Dr. Harry L. Shapiro, May 15, 1961, Box 2, DIHGR, UMA.

40. Mrs. Ernest J. Schrader to Sheldon Reed, January 16, 1951, Box 2, DIHGR, UMA.

41. S. C. Reed to Chas. S. Campbell, MD, March 14, 1968; Chas S. Campbell to S. C. Reed, February 23, 1968, and attachment, chap. 436, 1967 Sterilization Part, Sterilization for Social Protection, Box 2, DIHGR, UMA.

42. Seymour Kessler, "The Psychological Paradigm Shift in Genetic Counseling," *Social Biology* 27, no. 3 (1980): 167–85.

43. Reed, "Genetic Counseling (Women's Studies)."

44. S. C. Reed, "Etiology and Genetic Counseling in Mental Retardation" (1955), Box 4, DIHGR, UMA.

45. See Daniel J. Kevles, *In the Name of Eugenics: Genetics and the Uses of Human Heredity*, rev. ed. (Cambridge, MA: Harvard University Press, 1995).

46. S. C. Reed, "Human Factors in Genetic Counseling" (1967), Box 5, DIHGR, UMA.

47. I thank Igna Martin for her careful consultation of the Carl Rogers Papers at the Library of Congress and Beth Kaplan, archivist at the University of Minnesota Archives, for closely checking those papers.

48. See David J. Rothman, *Strangers at the Bedside: A History of How Law and Bioethics Transformed Medical Decision Making* (New Brunswick, NJ: Aldine Transaction, 2003).

49. Ibid.

50. See Jonsen, *Birth of Bioethics*.

51. Bruce Hilton, "Will the Baby Be Normal? . . . And What Is the Cost of Knowing?," *Hastings Center Report* 2, no. 3 (1972): 8–9.

52. See, for example, Marc Lappé, "Genetic Knowledge and the Concept of Health," *Hastings Center Report* 3, no. 4 (1973): 1–3. He continued to work on these issues; see Marc Lappé, "The Limits of Genetic Inquiry," *Hastings Center Report* 17, no. 4 (1987): 5–10.

53. Marc Lappé, "How Much Do We Want to Know about the Unborn?," *Hastings Center Report* 3, no. 1 (1973): 8–9.

54. Murray, "Deciphering Genetics."

55. See, for example, Erik Parens and Adrienne Asch, "The Disability Rights Critique of Prenatal Genetic Testing," Special Supplement, *Hastings Center Report* 29, no. 5 (1999): S1–S22.

56. See Mark M. Ravitch et al., "Closure of Duodenal, Gastric and Intestinal Stumps with Wire Staples: Experimental and Clinical Studies," *Annals of Surgery* 163, no. 4 (1966): 573–79.

57. Quoted in Armand Matheny Antommaria, "'Who Should Survive?: One of the Choices of Our Conscience': Mental Retardation and the History of Contemporary Bioethics," *Kennedy Institute of Ethics Journal* 16, no. 3 (2006): 206.

58. David C. Clark to Dr. Mary Ellen Avery, December 19, 1963, Folder: Speeches, Mongoloids with duodenal atresia, Speeches Mi-Wo, Box 119E2, Papers of Robert E. Cooke (REC), Alan Mason Chesney Medical Archives (AMCMA), Johns Hopkins Medical Institutions (JHMI), Baltimore, Maryland.

59. Case records, Folder: Speeches, Mongoloids with duodenal atresia, Speeches Mi-Wo, Box 119E2, REC, AMCMA, JHMI.

60. Richard L. Peck, "When Should the Patient Be Allowed to Die?," *Hospital Physician*, July 1972, 28–33, Folder: Right to Die, Box 119E2, REC, AMCMA, JHMI.

61. Robert E. Cooke to Mrs. Emily J. McKeown, November 10, 1972, Folder: Right to Die, Box 119E2, REC, AMCMA, JHMI.

62. Peck, "When Should the Patient Be Allowed to Die?"; also see Harry Nelson, "Hospital Let Retarded Baby Die, Film Shows," *Los Angeles Times*, October 17, 1971, A9.

63. Mrs. Marilyn Birkmeyer to Robert E. Cooke, November 1, 1972, Folder: Right to Die, Box 119E2, REC, AMCMA, JHMI.

64. Lorna M. Schroder to Dr. Robert E. Cooke, October 18, 1971, Folder: Right to Die, Box 119E2, REC, AMCMA, JHMI.

65. The film was accompanied by a discussion guide geared toward ethical decisions groups. See Discussion Guide, *Who Should Survive?*, A Film by the Joseph P. Kennedy, Jr. Foundation Medical Ethics Series, Joseph P. Kennedy, Jr. Foundation Film Services (1971).

66. Richard Heller to Theodore M. King, May 12, 1977, Box 504035, Howard W. Jones Jr. Papers, (HWJ), Folder: Prenatal (3 of 4), AMCMA, JHMI.

67. Tabitha M. Powledge and John Fletcher, "Guidelines for the Ethical, Social and Legal Issues in Prenatal Diagnosis," *New England Journal of Medicine* 300, no. 4 (1979): 171.

68. Arno Motulsky, "Brave New World? Ethical Issues in Prevention, Treatment and Research of Human Birth Defects," in Proceedings of the Fourth International Conference, ed. A. Motulsky & W. Lentz (Vienna, Austria, September 2–8, 1971), *Birth Defects, International Congress Series* 310 (1973): 319, 327. This article was republished the following year in *Science* as "Brave New World? Current Approaches to Prevention, Treatment, and Research of Genetic Diseases Raise Ethical Issues," *Science* 185 (1974): 653–63.

69. Motulsky, "Brave New World?," 318.

70. James R. Sorenson, Judith P. Swazey, and Norman A. Scotch, *Reproductive Pasts, Reproductive Futures: Genetic Counseling and Its Effectiveness*, Birth Defects Original Article Series 17, no. 4 (White Plains, NY: Alan R. Liss, 1981), 44.

71. Ibid., 42.

72. Jérôme Lejeune, "On the Nature of Men," *American Journal of Human Genetics* 22, no. 2 (1970): 128.

73. Melissa Richter to Messeur LeJeune, September 21, 1970, Record Group (RG) 2, Human Genetics Graduate Program Records (HGGPR), Sarah Lawrence College Archives (SLCA), Bronxville, New York.

74. Richter, Memorandum, January 12, 1972, RG 2, HGGPR, SLCA.

75. Course Description, Social Psychiatry, 1970–1971, RG 6, HGGPR, SLCA.

76. Seminar in Genetic Counseling, Fall 1973, and Issues in Clinical Genetics, 1979–1980, RG 6, HGGPR, SLCA; Joan Marks, e-mail message to author, June 24, 2008.

77. Seminar in Genetic Counseling, Fall 1974, RG 7, HGGPR, SLCA; Joan H. Marks, "2003 ASHG Award for Excellence in Human Genetics Education: The Importance of Genetic Counseling," *American Journal of Human Genetics* 74 (2004): 396.

78. Elsa Reich, interviewed by the author, August 16, 2011.

79. "Client-Centered Therapy of Personality and Therapy," Seminar in Genetic Counseling, Curriculum, 1977–1978; RG 6, HGGPR, SLCA; see Marvin Frankel and Lisbeth Sommerbeck, "Two Rogers and Congruence: The Emergence of Therapist-Centered Therapy and the Demise of Client-Centered Therapy," in *Embracing Nondirectivity: Reassessing Person-Centered Theory and Practice in the 21st Century*, ed. Brian E. Levitt (Ross-on-Wye, UK: PCCS Books, 2005), 40–61.

80. "Client-Centered Counseling: A Practicum," Seminar in Genetic Counseling, Curriculum, 1983–1984, RG 6, HGGPR, SLCA.

81. Marvin Frankel, e-mail message to author, July 17, 2011.

82. Ibid.

83. Marvin Frankel, e-mail message to author, June 1, 2010.

84. "Issues in Clinical Genetics," Curriculum, 1978–1979, RG 6, HGGPR, SLCA.

85. Human Genetics Program, 1972, RG 1; Additional Course Offering—Spring 1975, Developmental Biology, RG 6, HGGPR, SLC.

86. Kessler, "Psychological Paradigm Shift."

87. For a moving discussion of this see Seymour Kessler, "Notes and Reflections," chap. 13 in *Psyche and Helix: Psychological Aspects of Genetic Counseling*, ed. Robert G. Resta (New York: Wiley-Liss, 2000), 165–72.

88. Seymour Kessler, interviewed by the author, May 1, 2007.

89. Seymour Kessler, "Psychological Aspects of Genetic Counseling. XI. Nondirectiveness Revisited," *American Journal of Medical Genetics* 72 (1997): 164–71.

90. Seymour Kessler, "Psychological Aspects of Genetic Counseling. VII. Thoughts on Directiveness," *Journal of Genetic Counseling* 1, no. 1 (1992): 9–17.

91. Seymour Kessler, *Genetic Counseling: Psychological Dimensions* (New York: Academic Press, 1979).

92. Seymour Kessler, "Psychological Aspects of Genetic Counseling: Analysis of a Transcript," *American Journal of Medical Genetics* 8 (1981): 151.

93. Robert G. Resta, "Eugenics and Nondirectiveness in Genetic Counseling," *Journal of Genetic Counseling* 6, no. 2 (1997): 255–58; Angus Clarke, "Is Non-directive Genetic Counseling Possible?," *Lancet* 338 (1991): 998–1001; Mark Yarborough, Joan A. Scott, and Linda K. Dixon, "The Role of Beneficence in Genetic Counseling: Non-directive Counseling Reconsidered," *Theoretical Medicine* 10 (1989): 139–49; Susan Michie et al., "Nondirectiveness in Genetic Counseling: An Empirical Study," *American Journal of Human Genetics* 60 (1997): 40–47; Barbara A. Bernhardt, "Empirical Evidence That Genetic Counseling Is Directive: Where Do We Go from Here?," *American Journal of Human Genetics* 60 (1989): 17–20; Dianne M. Bartels et al., "Nondirectiveness in Genetic Counseling: A Survey of Practitioners," *American Journal of Medical Genetics* 72 (1997): 172–79.

94. Jon Weil, interviewed by the author, November 23, 2007.

95. Attachment C, "Bioethics," included with "Descriptive Materials regarding the Genetic Counseling Option for the Committee on Educational Policy," May 5, 1977, University of California at Berkeley Genetic Counseling Program (UCB), Personal Archive of Margie Goldstein (MG), Berkeley, California.

96. Bonnie LeRoy, interviewed by the author, January 30, 2008.

97. Minutes of the Minnesota Human Genetics League Board Meeting, May 2,

1991; Minutes of the Minnesota Human Genetics League Executive Meeting, March 5, 1991, Box 1, Minnesota Human Genetics League Records (MHGLR), UMA.

98. Minutes of the Minnesota Human Genetics League Board Meeting, May 2, 1991. Criticism of the Dight Institute's ability to keep pace with developments in human genetics was voiced in the early 1980s; see Richard A. King to V. Elving Anderson, March 17, 1983, Box 1, DIHGR, UMA.

99. Jon Weil, *Psychosocial Genetic Counseling* (Oxford: Oxford University Press, 2000), 124; also see Mary Terrell White, "Making Responsible Decisions: An Interpretive Ethic for Genetic Decisionmaking," *Hastings Center Report* 29, no. 1 (1999): 14–21.

100. David H. Smith et al., *Early Warning: Cases and Ethical Guidance for Presymptomatic Testing in Genetic Diseases* (Bloomington: Indiana University Press, 1998), 25.

101. Joan Marks, interviewed by the author, January 19, 2007.

102. Barbara Biesecker, interviewed by the author, February 1, 2007.

103. Luba Djurdjinovic, interviewed by the author, January 25, 2008.

104. Weil, *Psychosocial Genetic Counseling*, 125.

105. Arno Motulsky, interviewed by the author, March 6, 2008; American Society of Human Genetics, Conversations in Genetics, "Talking with Arno Motulsky" (on CD) (Betheseda, MD: Genetics Society of America, 2003).

106. See, for example, V. E. Headings, "Revisiting Foundations of Autonomy and Beneficence in Genetic Counseling," *Genetic Counseling* 8, no. 4 (1997): 291–94; Jan Hodgson and Merle Spriggs, "A Practical Account of Autonomy: Why Genetic Counseling Is Especially Well Suited to the Facilitation of Informed Autonomous Decision Making," *Journal of Genetic Counseling* 14, no. 2 (2005): 89–97; and Mary Terrell White, "'Respect for Autonomy' in Genetic Counseling: An Analysis and a Proposal," *Journal of Genetic Counseling* 6, no. 3 (1997): 297–313.

CHAPTER SEVEN: *Prenatal Diagnosis*

1. Virginia Corson, interviewed by the author, January 30, 2007; Alan L. Otten, "Parental Agony: How Counselors Guide Couples When Science Spots Genetic Risks," *Wall Street Journal*, March 8, 1989, A1, A8, Subject Files: Genetic Diseases (GD); "Geneticist Predicts Defects," *Messenger*, February 27, 1980, 1–3, Biographical File: Virginia Corson (VC), Alan Mason Chesney Medical Archives (AMCMA), The Johns Hopkins Medical Institutions (JHMI), Baltimore, Maryland.

2. The group was the Garrod-Galton Society; for a fuller analysis of its significance to the development of medical genetics in America see Nathaniel Comfort, *The Science of Human Perfection: Heredity, Health, and Human Improvement in American Biomedicine* (New Haven, CT: Yale University Press, 2012).

3. "The History of Human Genetics in the Johns Hopkins University," Box 436833458, John Littlefield Papers (JL), Folder: Genetics Unit, AMCMA, JHMI; Lisa Harris, "In Vitro Fertilization in the United States: A Clinical and Cultural History" (PhD dissertation, University of Michigan, 2006); Robin Marantz Henig, *Pandora's Baby: How the First Test Tube Babies Sparked the Reproductive Revolution* (New York: Houghton Mifflin, 2004).

4. Michael M. Kaback et al., "Approaches to the Control and Prevention of

Tay Sachs Disease," in *Progress in Medical Genetics*, vol. 10, ed. A. Steinberg and A. Bearn (New York: Grune and Stratton, 1974), 103–34; Howard Markel, "Scientific Advances and Social Risks: Historical Perspectives of Genetic Screening Programs for Sickle Cell Disease, Tay Sachs Disease, Neural Tube Defects, and Down Syndrome, 1970–1997," in *Promoting Safe and Effective Genetic Testing in the United States. Final Report of the Task Force on Genetic Screening (Appendix 6)*, ed. Neil A. Holtzman and Michael S. Watson (Bethesda, MD: NIH-DOE Working Group on Ethical, Legal and Social Implications of Human Genome Research, 1997), www.genome.gov/10002401; Keith Wailoo and Stephen Pemberton, *The Troubled Dream of Genetic Medicine: Disease and Ethnicity in Tay-Sachs, Cystic Fibrosis, and Sickle Cell Disease* (Baltimore: Johns Hopkins University Press, 2006).

5. Haig W. Kazazian Jr., interviewed by Nathaniel Comfort, July 13, 2006, Oral History of Human Genetics Project, accessible at http://ohhgp.pendari.com/Links.aspx.

6. John Littlefield to Dr. Theodore M. King, September 14, 1977, Box 431969216/509130, Victor McKusick Papers (VM), Folder: Prenatal Diagnostic Center; Richard H. Heller to Haig Kazazian, May 29, 1978, Howard W. Jones Jr. Papers (HWJ), Folder: Prenatal Diagnostic Center (3 of 4), AMCMA, JHMI.

7. Corson, interview; Judy Minkove, "The Gene Gurus: Genetic Counselors Help Us Grasp Our Biological Destiny," *Dome* 56, no. 4 (2005): 5–6, Subject Files: Genetic Counseling Clinics (GCC), AMCMA, JHMI.

8. Jennifer A. Bubb and Anne L. Matthews, "What's New in Prenatal Screening and Diagnosis?," *Primary Care: Clinics in Office Practice* 31, no. 3 (2004): 561–82.

9. Corson, interview; Otten, "Parental Agony."

10. See Robert G. Resta, "Historical Aspects of Genetic Counseling: Why Was Maternal Age 35 Chosen as the Cut-Off for Offering Amniocentesis," *Medicina nei Secoli Arte e Scienza* 14, no. 3 (2002): 793–811.

11. Luba Djurdjinovic, interviewed by the author, September 22, 2011.

12. Corson, interview.

13. See, for example, Deborah Kaplan, "Prenatal Screening and Its Impact on Persons with Disabilities," *Clinical Obstetrics and Gynecology* 36, no. 3 (1993): 605–12; Rayna Rapp, *Testing Women, Testing the Fetus: The Social Impact of Amniocentesis in America* (New York: Routledge, 2003).

14. Sonia Mateu Suter, "The Routinization of Prenatal Testing," *American Journal of Law & Medicine* 28 (2002): 233–70.

15. Memorandum for the Record, November 15, 1968, Box 504035, HWJ, Folder: Prenatal (2 of 4), AMCMA, JHMI.

16. "Prenatal Birth Defects Diagnostic Center at the Johns Hopkins Hospital," Box 504035, HWJ, Folder: Prenatal (2 of 4), AMCMA, JHMI.

17. Leslie J. Reagan, *Dangerous Pregnancies: Mothers, Disabilities, and Abortion in Modern America* (Berkeley: University of California Press, 2010); National Foundation of Infant Paralysis, *Expanded Program* (1958), 15–29, Medical Programs Records (MPR), Series 6: Expanded Program, Box 6, March of Dimes Archives (MDA), White Plains, New York.

18. In 1972, the National Foundation for Infant Paralysis did not want to get embroiled in the issue of abortion, stating that "legal abortion is outside the Founda-

tion's purview" and strictly a parental decision. See Memorandum, George P. Voss to Chairman of Chapters with a population of 25,000 and over, March 14, 1972, MPR, Series 3: Birth Defects, Box 3, Folder: Amniocentesis (1972), MDA.

19. Form Letter, Medical Department, the National Foundation–March of Dimes, October 1, 1969; Memorandum, Re: NF-MOD visibility in genetics, Dorothy Davis to Dr. Virginia Apgar, June 16, 1970, MPR, Series 3: Birth Defects, Box 3, MDA.

20. Richard H. Heller to John Littlefield, November 21, 1973, Box 436833456, JL, Folder: Prenatal, AMCMA, JHMI.

21. "Prenatal Birth Defects Diagnostic Center."

22. Marc Lappé, "How Much Do We Want to Know about the Unborn?," *Hastings Center Report* 3, no. 1 (1973): 8–9.

23. Sherman Elias, Joe Leigh Simpson, and Allan T. Bombard, "Amniocentesis and Fetal Blood Sampling," chap. 2 in *Genetic Disorders and the Fetus: Diagnosis, Prevention, and Treatment*, ed. Aubrey Milunsky, 4th ed. (Baltimore: Johns Hopkins University Press, 1998), 53–82.

24. Robert G. Resta, "The First Prenatal Diagnosis of a Fetal Abnormality," *Journal of Genetic Counseling* 6, no. 1 (1997): 81–83.

25. Resta, "Historical Aspects of Genetic Counseling"; Ruth Schwartz Cowan, "Women's Roles in the History of Amniocentesis and Chorionic Villi Sampling," in *Women and Prenatal Testing: Facing the Challenges of Genetic Technology*, ed. Karen Rothenberg and Elizabeth Thomson (Columbus: Ohio State University Press, 1994), 35–48; Jean Marie Brady, "Discussion of Amniotic Cell Cultures in Prenatal Diagnosis for Genetic Counseling," *American Journal of Medical Technology* 37, no. 11 (1971): 428–33.

26. Comfort, *Science of Human Perfection*; Ruth Schwartz Cowan, *Heredity and Hope: The Case for Genetic Screening* (Cambridge, MA: Harvard University Press, 2008).

27. Daniel J. Kevles, *In the Name of Eugenics: Genetics and the Uses of Human Heredity*, 2nd ed., (Cambridge, MA: Harvard University Press, 1995).

28. Fiona Alice Miller, "A Blueprint for Defining Health: Making Medical Genetics in Canada, c. 1935–1975" (PhD dissertation, York University, February 2000).

29. See Cowan, *Heredity and Hope*, chap. 3.

30. Albert B. Gerbie and Arnold A. Shkolnik, "Ultrasound Prior to Amniocentesis for Genetic Counseling," *Obstetrics and Gynecology* 46, no. 6 (1975): 716–19.

31. Mark W. Steele and W. Roy Breg Jr., "Chromosome Analysis of Human Amniotic-Fluid Cells," *Lancet* 1 (1966): 385.

32. "Prenatal Birth Defects Diagnostic Center."

33. Charles J. Epstein et al., "Prenatal Detection of Genetic Disorders," *American Journal of Human Genetics* 24 (1972): 214–26; Lillian Y. F. Hsu et al., "Results and Pitfalls in Prenatal Cytogenetic Diagnosis," *Journal of Medical Genetics* 10 (1973): 112–19; Aubrey Milunsky and Leonard Atkins, "Prenatal Diagnosis of Genetic Disorders: An Analysis of Experience with 600 Patients," *Journal of the American Medical Association* 230, no. 2 (1974): 232–35; Gerald H. Prescott et al., "A Prenatal Diagnosis Clinic: An Initial Report," *American Journal of Obstetrics and Gynecology* 116, no. 7 (1973): 942–48.

34. Kurt Hirschhorn, interviewed by the author, March 9, 2009; also see Kurt

Hirschhorn, "The Role and Hazards of Amniocentesis," *Annals of the New York Academy of Medicine* 240 (1975): 117–20.

35. Henry L. Nadler and Albert B. Gerbie, "Role of Amniocentesis in the Intrauterine Detection of Genetic Disorders," *New England Journal of Medicine* 282, no. 11 (1970): 599.

36. Ibid.

37. Epstein et al., "Prenatal Detection of Genetic Disorders," 224.

38. Joseph R. Hixson, "Forecasts from the Womb," *New York Times*, January 19, 1967, 56; Robert W. Stock, "Will the Baby Be Normal? The Genetic Counselor Tries to Find the Answer by Translating the Biological Revolution into Human Terms," *New York Times*, March 23, 1969, SM25.

39. Jane E. Brody, "Prenatal Diagnosis Is Reducing Risk of Birth Defects," *New York Times*, June 3, 1971, 41.

40. Gilbert S. Omenn, "Prenatal Diagnosis of Genetic Disorders," *Science* 200, no. 26 (1978): 952–58; Charles J. Epstein and Mitchell S. Golbus, "The Prenatal Diagnosis of Genetic Disorders," *Annual Reviews of Medicine* 29 (1978): 117–28.

41. Leslie J. Reagan, *When Abortion Was a Crime: Women, Medicine, and Law in the United States, 1867–1973* (Berkeley: University of California Press, 1997).

42. See, for example, Walter Sullivan, "Wider Detection of Prenatal Flaws Expected to Spur Abortions," *New York Times*, June 13, 1970, 11.

43. Richard H. Heller, MD, Curriculum Vitae, included in Questionnaire, March of Dimes 1973–1974, Box 504305, HWJ, Folder: Prenatal (4 of 4), AMCMA, JHMI.

44. Richard H. Heller, letter to the editor, most likely the *Baltimore Sun*, November 3, 1971, Biographical Files, Folder: Richard Heller (RH), AMCMA, JHMI.

45. World Health Organization, "Genetic Counseling: Third Report of the WHO Expert Committee on Human Genetics," *World Health Organization Technical Report Series*, no. 416 (Geneva: World Health Organization, 1969), 10.

46. "Agenda, Amniocentesis Registry Meeting, July 12, 1973," and accomanying charts, Box 436833456, JL, Folder: Prenatal, AMCMA, JHMI.

47. The NICHD National Registry for Amniocentesis Study Group, "Midtrimester Amnioctenesis for Prenatal Diagnosis," *JAMA* 236, no. 13 (1976): 1475. The article indicates that the Eunice Kennedy Shriver Center in Boston and the University of California at San Diego joined the study later.

48. M. d'A. Crawfurd et al., "Early Prenatal Diagnosis of Hurler's Syndrome with Termination of Pregnancy and Confirmatory Findings on the Fetus," *Journal of Medical Genetics* 10 (1973): 144–53; Helmut G. Schrott, Laurence Karp, and Gilbert S. Omenn, "Prenatal Prediction in Myotonic Dystrophy: Guidelines for Genetic Counseling," *Clinical Genetics* 4 (1973): 38–45; Roscoe O. Brady, B. William Uhlendorf, and Cecil B. Jacobson, "Fabry's Disease: Antenatal Detection," *Science* 172, no. 3979 (1971): 174–75.

49. Cowan, *Heredity and Hope*, 102; Centers for Disease Control and Prevention, "Chorionic Villus Sampling and Amniocentesis: Recommendations for Prenatal Counseling," *Morbidity and Mortality Weekly Report* 44 (July 21, 1995): 2; Tabitha M. Powledge, "Prenatal Diagnosis: New Techniques, New Questions," *Hastings Center Report* 9, no. 3 (1979): 16–17.

50. Tamsen M. Caruso, Marie-Noel Westgate, and Lewis B. Holmes, "Impact of Prenatal Screening on the Birth Status of Fetuses with Down Syndrome at an Urban Hospital, 1972–1994," *Genetics in Medicine* 1, no. 1 (1998): 22–28.

51. Mitchell S. Golbus et al., "Prenatal Genetic Diagnosis in 3000 Amniocenteses," *New England Journal of Medicine* 300, no. 4 (1979): 157–63.

52. Resta, "Historical Aspects of Genetic Counseling."

53. Chart, Growth of the Prenatal Diagnostic Center of the Johns Hopkins Hospital, Box 504035, HWJ, Folder: Prenatal (4 of 4), AMCMA, JHMI.

54. Statistical Summary, Prenatal Diagnostic Center—Johns Hopkins Hospital, 1976, included in Richard Heller to Dr. Benjamin White, March 29, 1977, Box 405035, HWJ, Folder: Prenatal (4 of 4), AMCMA, JHMI.

55. Minutes of the Prenatal Diagnostic Center, Medical Advisory Board, April 2, 1973, and October 7, 1974, Box 504035, HWJ, Folder: Prenatal (1 of 4), AMCMA, JHMI.

56. Minutes, Prenatal Diagnostic Center, Meeting of the Medical Advisory Board, January 15, 1973, Box 504035, HWJ, Folder: Prenatal (2 of 4), AMCMA, JHMI.

57. Minutes of the Meeting of the Prenatal Diagnostic Center, Medical Advisory Board, October 15, 1973; Minutes of the Medical Advisory Board, PDC, October 15, 1973, Box 436833458, JL, Folder: Prenatal, AMCMA, JHMI.

58. Richard Heller to Dr. John Littlefield, January 25, 1974, Box 436833458, JL, Folder: Prenatal, AMCMA, JHMI.

59. Richard Heller to Dr. Howard W. Jones, March 10, 1977, Box 504035, HWJ, Folder: Prenatal (3 of 4), AMCMA, JHMI.

60. Minutes of the Meeting of the Prenatal Diagnostic Center, Medical Advisory Board, October 7, 1974, Box 504035, HWJ, Folder: Prenatal (1 of 4), AMCMA, JHMI.

61. Ibid.

62. Minutes of the Medical Advisory Board Meeting of the Prenatal Diagnostic Clinic, January 24, 1978, Box 504035, HWJ, Folder: Prenatal (3 of 4), AMCMA, JHMI.

63. Ibid.

64. Ibid.

65. Richard H. Heller to John Littlefield, November 21, 1973, Box 436833456, JL, Folder: Prenatal; Questionnaire, March of Dimes, 1973–1974, Box 504305, HWJ, Folder: Prenatal (4 of 4), AMCMA, JHMI.

66. Jane E. Brody, "Genetics Clinics Predict Defects," *New York Times*, February 2, 1969, 76; also see her "Medicine to Forecast Birth Defects," *New York Times*, May 25, 1969, E8.

67. Questionnaire, March of Dimes, 1973–1974, Box 504305, HWJ, Folder: Prenatal (4 of 4), AMCMA, JHMI.

68. "Life Quality Paramount, Seminar Topic Here," *Daily Times* (Salisbury, MD), November 22, 1976, n.p., RH, AMCMA, JHMI.

69. Obituary, *Baltimore Sun*, February 9, 1982, D6, RH, AMCMA, JHMI.

70. Ronald Conley and Aubrey Milunsky, "The Economics of Prenatal Genetic Diagnosis," chap. 20 in *The Prevention of Genetic Disease and Mental Retardation*, ed. Aubrey Milunsky (Philadelphia: W. B. Saunders, 1975), 442–55.

71. "Prenatal Birth Defects Diagnostic Center."

72. Neil A. Holtzman to Benjamin White, July 16, 1973, Box 436833458, JL, Folder: Genetics Unit, AMCMA, JHMI.

73. Gerald H. Prescott et al., "A Prenatal Diagnosis Clinic: An Initial Report," *American Journal of Obstetrics and Gynecology* 116, no. 7 (1973): 945, 948.

74. Resta, "Historical Aspects of Genetic Counseling," 797; Susan P. Pauker and Stephen G. Pauker, "Prenatal Diagnosis—Why Is 35 a Magic Number?," *New England Journal of Medicine* 330 (1994): 1151–52.

75. David S. Newberger, "Down Syndrome: Prenatal Risk Assessment and Diagnosis," *American Family Physician*, www.aafp.org/afp/20000815/825.html.

76. Cowan, "Women's Roles in the History of Amniocentesis," 35–48; George J. Annas and Brian Coyne, "'Fitness' for Birth and Reproduction: Legal Implications of Genetic Screening," *Family Law Quarterly* 9 (1975): 463–89; Captain Jeffrey L. Grundtisch, "Legal Liability in Genetic Counseling and Testing," *Air Force Law Review* 21, no. 3 (1979): 462–72; Aubrey Milunsky and Philip Reilly, "The 'New' Genetics: Emerging Medicolegal Issues in the Prenatal Diagnosis of Hereditary Disorders," *American Journal of Law and Medicine* 1 (1975): 71–88.

77. Proposal, Statewide Screening for Spina Bifida by α-Fetoprotein, Box 504035, HWJ, Folder: Prenatal (4 of 4), AMCMA, JHMI.

78. Minutes of the Meeting of the Prenatal Diagnostic Center, Medical Advisory Board, November 5, 1973, Box 436833458, JL, Folder: Prenatal, AMCMA, JHMI.

79. Richard Heller to Dr. H. Lorrin Lau, July 15, 1976, Box 504035, HWJ, Folder: Prenatal (4 of 4), AMCMA, JHMI.

80. Aubrey Milunsky and Elliott Alpert, "Sounding Board: Antenatal Diagnosis, Alpha Fetoprotein and the FDA," *New England Journal of Medicine* 295, no. 3 (1976): 169.

81. Proposal, Statewide Screening for Spina Bifida.

82. See George J. Annas, "At Law: Is a Genetic Screening Test Ready When the Lawyers Say It Is?," *Hastings Center Report* 15, no. 6 (1985): 16–18. In 1986 California began to offer MSAPF to all patients accessing the state's prenatal services. See Carole H. Browner and Nancy Ann Press, "The Normalization of Prenatal Diagnostic Screening," chap. 17 in *Conceiving the New World Order: The Global Politics of Reproduction*, ed. Faye D. Ginsburg and Rayna Rapp (Berkeley: University of California Press, 1995), 307–22; Nancy Press and C. H. Browner, "Why Women Say Yes to Prenatal Diagnosis," *Social Science & Medicine* 45, no. 7 (1997): 979–89.

83. "Prenatal Birth Defects Diagnostic Center."

84. "'Dial the Doctor' for Info on Genetic Counseling," *Salisbury Advertiser*, November 14, 1974, 9, RH, AMCMA, JHMI.

85. Kazazian, interview.

86. Mt. Sinai Amniocentesis, Record Group (RG) 6, Curriculum, 1975–1976, Fieldwork, Human Genetics Graduate Program Records (HGGPR), Sarah Lawrence College Archives (SLCA), Bronxville, New York.

87. Djurdjinovic, interview.

88. Elsa Reich, interviewed by the author, August 16, 2011.

89. Minutes of the Meeting of the Prenatal Diagnostic Center, Medical Advisory Board, May 17, 1976, Box 426833458, JL, Folder: Prenatal, AMCMA, JHMI.

90. Haig H. Kazazian Jr., "A Medical View," *Hastings Center Report* 10, no. 1 (1980): 17.

91. Ibid., 18.

92. Barbara Katz Rothman, *The Tentative Pregnancy: How Amniocentesis Changes the Experience of Motherhood* (New York: W. W. Norton, 1986).

93. Also see Mitchell S. Golbus et al., "Intrauterine Diagnosis of Genetic Defects: Results, Problems, and Follow-Up of One Hundred Cases in a Prenatal Genetic Detection Center," *American Journal of Obstetrics and Gynecology* 118, no. 7 (1974): 897–905; Bruce D. Blumberg, Mitchell S. Golbus, and Karl H. Hanson, "The Psychological Sequelae of Abortion Performed for a Genetic Indication," *American Journal of Obstetrics and Gynecology* 122, no. 7 (1975): 799–808; Bruce D. Blumberg, Mitchell S. Golbus, and Karl H. Hanson, "The Psychological Sequale of Abortion Performed for a Genetic Indication," *Obstetrics and Gynecology* 122, no. 7 (1975): 799–808.

94. Epstein et al., "Prenatal Detection of Genetic Disorders," 220.

95. Sara C. Finley et al., "Participants' Reactions to Amniocentesis and Prenatal Genetics Studies," *JAMA* 238, no. 22 (1977): 2377–79.

96. Dorothy C. Wertz and John C. Fletcher, "A Critique of Some Feminist Challenges to Prenatal Diagnosis," *Journal of Women's Health* 2, no. 2 (1993): 183.

97. Susan Markens, C. H. Browner, and H. Mabel Preloran, "'I'm Not the One They're Sticking the Needle Into': Latino Couples, Fetal Diagnosis, and the Discourse of Reproductive Rights," *Gender & Society* 17, no. 3 (2003): 465.

98. Cowan, *Heredity and Hope*.

99. C. H. Browner and H. Mabel Preloran, "Culture and Communication in the Cultural Realm of Fetal Diagnosis: Unique Considerations for Latino Patients," in *Genetic Testing: Care, Consent, and Liability*, ed. Neil F. Sharpe and Ronald F. Carter (New York: Wiley-Liss, 2006), 33.

100. Ibid., 44.

101. See Seymour Kessler, "Notes and Reflections," chap. 13 in *Psyche and Helix: Psychological Aspects of Genetic Counseling*, ed. Robert G. Resta (New York: Wiley-Liss, 2000), 165–66.

102. Gary Frohlich, Michael Begleiter, June Peters, and Bonnie Baty, interviewed by the author, December 18, 2007.

103. Robert Resta, interviewed by the author, March 20, 2007.

104. Rapp, *Testing Women, Testing the Fetus*; Rothman, *Tentative Pregnancy*; Neil A. Holtzman, *Proceed with Caution: Predicting Genetic Risks in the Recombinant DNA Era* (Baltimore: Johns Hopkins University Press, 1989).

105. Reich, interview.

CONCLUSION

1. Barton Childs, interviewed by Andrea Maestrejuan, December 12, 2001, Oral History of Human Genetics Project, UCLA and Johns Hopkins, http://ohhgp.pendari.com/Links.aspx.

2. Misha Angrist, *Here Is a Human Being: At the Dawn of Personal Genomics* (New York: HarperCollins, 2010).

3. Luba Djurdijnovic, interviewed by the author, January 25, 2008.

4. Wendy Uhlmann, interviewed by the author, December 19, 2007; Alan E. Guttmacher, Jean Jenkins, and Wendy R. Uhlmann, "Genomic Medicine: Who Will Practice It? A Call to Open Arms," *American Journal of Medical Genetics* 106 (2001): 216–22.

5. Catherine Reiser and Joan Burns, interviewed by the author, October 22, 2010.

6. See Angrist, *Here Is a Human Being*.

7. Andrew Pollack, "A Less Risky Down Test Lifts Hopes," *New York Times*, October 17, 2011, B1.

8. E-mail correspondence with Robert Resta, October 18, 2011.

9. Amy Harmon, "Love You, K2a2a, Whoever You Are," *New York Times*, January 22, 2006, www.nytimes.com/2006/01/22/weekinreview/22harmon.html?page wanted=all.

10. See Agrist, *Here Is a Human Being*; and Shobita Parthasarathy, "Assessing the Social Impact of Direct-to-Consumer Genetic Testing: Understanding Sociotechnical Architectures," *Genetics in Medicine* 12, no. 9 (2010): 544–47.

11. Kathryn T. Hock et al., "Direct-to-Consumer Genetic Testing: An Assessment of Genetic Counselors' Knowledge and Beliefs," *Genetics in Medicine* 13, no. 4 (2011): 325–32; United States Government Accountability Office, "Direct-to-Consumer Genetic Tests: Misleading Test Results Are Further Complicated by Deceptive Marketing and Other Questionable Practices," GAO-10-847, July 22, 2010, www.gao.gov/products/GAO-10-847T; Shobita Parthasarathy, *Building Genetic Medicine: Breast Cancer, Technology, and the Comparative Politics of Health Care* (Boston: MIT Press, 2007); Cheryl Berg and Kelly Fryer-Edwards, "The Ethical Challenges of Direct-to-Consumer Genetic Testing," *Journal of Business Ethics* 77 (2008): 17–32.

12. See Berg and Fryer-Edwards, "Ethical Challenges"; and "ASHG Statement on Direct-to-Consumer Genetic Testing in the United States," *American Journal of Human Genetics* 81 (2007): 635–37.

13. Beverly Yashar, interviewed by the author, December 15, 2010.

14. Angrist, *Here Is a Human Being*.

15. Karen Heller, interviewed by the author, October 28, 2010.

16. Heller, interview.

17. 2010 *Professional Status Survey: Salary and Benefits* (Chicago: National Society of Genetic Counselors, 2010), 18–19, available at www.nsgc.org.

18. Barbara Biesecker, interviewed by the author, February 1, 2007.

19. Kelly Ormond, "NSGC Foundations—Then, Now, Tomorrow," *Journal of Genetic Counseling* 14, no. 2 (2005): 85–88.

20. Kelly Ormond, interviewed by the author, September 27, 2011.

21. Caroline Lieber, interviewed by the author, November 5, 2010.

22. This toolkit is available at www.geneticcounselingtoolkit.com/about.htm.

prenatal screening *(cont.)*
and bias against Down syndrome,
99–100; and cost analysis, 163–64;
and eugenics, 167; and risk, 4, 31; and
training, 168
prenatal testing, 7, 9, 15, 24, 100, 148,
168–69; and bias against disabilities,
12, 76, 101; and determination of sex,
156, 157–58, 165–66; and Djurdjinovic, 164–65; and ethics, 133; and
eugenics, 98, 161, 167; growth of, 147;
and Johns Hopkins Hospital, 160;
and Lappé, 133; new tests for, 171; and
E. Reich, 165. *See also* amniocentesis
prevention, 10, 14, 17, 18, 76, 97, 98,
168; and abortion, 97; and Allan, 61;
and Childs, 25; and cost-analysis, 25,
162; and disabilities, 76, 85, 88; and
Lappé, 124; as luxury for wealthy, 12;
and Marks and Hendin, 91, 93; and
Murray, 85; possibilities for, 23; and
S. Reed, 43, 88, 97; and Richter, 109,
137; and risk information, 16; as unattainable goal, 18
Protestants, 66, 156
psychiatry, 11, 76
psychoanalysis, 125, 126, 127, 138
psychology, 14, 18, 22, 138, 165
psychosocial dynamic, 6, 14, 140, 148
psychotherapy, 125, 126
Puñales-Morejon, Diana, 39–40

Rabinow, Paul, 89
race, 3, 4–5, 9, 19, 53–74, 136, 167–68;
and adoption, 63–72; and diversity
training, 74; and Dobzhansky and
Dunn, 56–57; and Down syndrome,
89–90, 91; and Draper, 58–61; and
eugenics, 56; and medical genetics,
73; and Neel, 59, 66–67, 69, 71; and
physical traits, 64, 69; and population, 55–58; and S. Reed, 53–55, 59,
62–66, 69, 71, 129; and risk assessment, 42; and skin color, 53–54, 63,
64, 66–67, 68, 69; and superiority
and inferiority, 73; and University of
Michigan Heredity Clinic, 66–72. *See
also* eugenics
Rank, Otto, 126
Raushenbush, Esther, 104, 110

Reardon, Jenny, 56
recessive conditions, 40
recessive disorders, 42
Reed, Elizabeth, 35, 72, 80, 81, 87
Reed, Sheldon, 43, 45, 49, 72, 87, 88, 97,
129; and adoption, 63–66, 69, 71; and
autonomy, 125, 129–32; breast cancer
study by, 41; and client evasions, 46;
clients of, 34, 129, 131; and client's
reactions to risk, 47; and client vs.
patient, 6, 125, 131–32, 143; and
compassion, 21, 22, 88; and Dight
Institute, 77, 79, 129; and Down
syndrome, 40–41; and Draper, 62;
and educational level of counselors,
111; equivocal approaches of, 129–32;
and eugenics, 19–21, 36, 58, 62–63,
79, 80, 129, 131; and Faribault State
Home, 80; and genetic risk, 38, 39,
40–41; and Goethe, 62–63; as illustrative case, 35; and individuality, 131;
and intelligence, 58; and LeRoy, 142;
and Ludmerer, 77–78; and mental
retardation, 36, 79–83, 85, 87, 88, 97;
motivation of, 77–78; and Neel, 59;
and nondirectiveness, 132, 143; and
organizations for disabled groups, 89;
and Population Council, 72; and populations, 57, 58, 72; and race, 53–55,
59, 62–66, 69, 71, 129; and Richter,
104, 107, 109; and risk, 43, 45; and
Rogers, 129, 132; and sterilization,
131; and term genetic counseling, 4,
19–20, 77, 129
Reich, Annie, 138
Reich, Elsa, 138, 165, 169
Reiser, Catherine, 121, 171
reproduction, 2, 22, 109, 129–30, 148;
and autonomy, 6, 148, 166; and
choice, 76–77, 109, 132; decisions
about, 133, 137; and rights, 125, 167.
See also mothers; parents
Resta, Robert, 24, 26, 34, 162, 168–69,
171
Riccardi, Vincent M., 118
Richter, Melissa, 102, 104–12, 113, 114,
116, 118, 123; and ethics, 137–38;
orientation of, 105–9; and Sarah Lawrence context and location, 109–12
Right to Life, 153, 155